Surveying

Surveying

Editor

Marian Sewick

Surveying

Edited by **Marian Sewick**

Printed in 2017

ISBN: 978-1-68117-175-3
Library of Congress Control Number: 2015951146

© 2016 by
SCITUS Academics LLC,
616, Corporate Way, Suite 2, 4766,
Valley Cottage, NY 10989

www.scitusacademics.com

Preface

Surveying or land surveying is the technique, profession, and science of determining the terrestrial or three-dimensional position of points and the distances and angles between them. A land surveying professional is called a land surveyor.

Surveying is as old as the human civilization. The art of surveying and map drawing has been in practice since the cultural evolution of mankind. The earliest methods of surveys were made in connection with land surveying for the purpose of establishing boundaries of lands, but with the passage of time, an urge was felt to implement its application in many other avenues as well. The main development of surveying took place in the nineteenth century after the invention of telescope, magnetic compass, levelling instruments and theodolites. For the purpose of engineering projects such as roads, railways, canals, water supply, reservoirs, dams, building, bridges, flyovers, etc., extensive surveying is inevitable for proper establishment and allocation of the jobsite. The success of any engineering project is highly dependent on the accurate and complete survey work.

This book contributes to enhance the basic knowledge of the subject for the civil engineering students. The book has been prepared in such a way that it highlights every aspect of the subject—from the basic measurement technique by chains and tapes to the advanced features like application of EDM instruments, photogrammetry and remote sensing. Organised into 25 chapters this book highlights all the elements of surveying systematically. The chapters are arranged in a logical sequence in order to maintain the continuity. The theories are explained in a simple and lucid language along with the solved examples and problems.

The book explains the theories behind modem optical instruments like Electronic Distance Measurements (EDM), and Total stations, which are invented to give accurate measurements. The book shows how photogrammetric surveying is making a new headway with aircrafts, satellites and modem cameras. It also highlights the ways through which surveying is extended to the deep sea, and extra terrestrial space. Most importantly, it discusses how surveying principles have been used in remote sensing, rocket tracks, missiles and space vehicles.

Table of Contents

CHAPTER 3   **Advancing of Land Surface Temperature Retrieval Using Extreme Learning Machine and Spatio-Temporal Adaptive Data Fusion Algorithm........................ 59**

CHAPTER 4   **GIS Applied to Integrated Coastal Zone and Ocean Management: Mapping, Change Detection and Spatial Modeling for Coastal Management in Southern Brazil... 85**

CHAPTER 1

Integration of UAV-Based Photogrammetry and Terrestrial Laser Scanning for the Three-Dimensional Mapping and Monitoring of Open-Pit Mine Areas

Xiaohua Tong, Xiangfeng Liu, Peng Chen, Shijie Liu, Kuifeng Luan, Lingyun Li, Shuang Liu, Xianglei Liu, Huan Xie, Yanmin Jin and Zhonghua Hong

College of Surveying and Geo-Informatics, Tongji University, 1239 Siping Road, Shanghai 200092, China;

ABSTRACT

This paper presents a practical framework for the integration of unmanned aerial vehicle (UAV) based photogrammetry and terrestrial laser scanning (TLS) with application to open-pit mine areas, which includes UAV image and TLS point cloud acquisition, image and cloud point processing and integration, object-oriented classification and three-dimensional (3D) mapping and monitoring of open-pit mine areas. The proposed framework was tested in three open-pit mine areas in southwestern China. (1) With respect to extracting the conjugate points of the stereo pair of UAV images and those points between TLS point clouds and UAV images, some feature points were first extracted by the scale-invariant feature transform (SIFT) operator and the outliers were identified and therefore eliminated by the RANdom SAmple Consensus (RANSAC) approach; (2) With respect to improving the accuracy of geo-positioning based on UAV imagery, the ground control points (GCPs) surveyed from global positioning systems (GPS) and the feature points extracted from TLS were integrated in the bundle adjustment, and three scenarios were designed and compared; (3) With respect to monitoring and mapping the mine areas for land reclamation, an object-based image analysis approach was used for the classification of the accuracy improved UAV ortho-image. The experimental results show that by introduction of TLS derived point clouds as GCPs, the accuracy of geo-positioning based on UAV imagery can be improved. At the same time, the accuracy of geo-positioning based on GCPs form the TLS derived point clouds is close to that based on GCPs from the GPS survey. The results also show that the TLS derived point clouds can be used as GCPs in areas such as in mountainous or high-risk environments where it is difficult to conduct a GPS survey. The proposed framework achieved a decimeter-level accuracy for the generated digital surface model (DSM) and digital orthophoto map (DOM), and an overall accuracy of 90.67% for classification of the land covers in the open-pit mine.

INTRODUCTION

With the advent and availability of accurate and miniature global navigation satellite systems (GNSS) and inertial measurement units (IMUs), together with the availability of quality consumer-grade digital cameras and other miniature sensors, the unmanned aerial vehicle (UAV) technique has been developing rapidly in the civilian community [1,2,3]. The term UAV is defined as "uninhabited and reusable motorized aerial vehicles" [4], which can be remotely controlled by means of autonomous, semi-autonomous or manual operations. The UAV systems, which are based on multiple low-cost and conventional platforms [5,6,7,8] and are equipped with multiple flexible and efficient sensors [9,10,11], are able to acquire high-resolution images for photogrammetric and remote sensing applications.

In general, three types of space-borne and airborne remote sensing techniques are used for the acquisition of high-resolution images of large areas: satellites, manned aircraft, and UAVs. Satellites are often restricted by the running cycle, height orbit and cloud cover, and manned aircraft are affected by airspace limits and weather [12,13]. UAV-based remote sensing systems have both the common characteristics of aerial remote sensing and their own unique features. Compared to manned aircraft systems, UAVs, which need to obtain flight permission from the civilian aviation authority, can be deployed quickly and repeatedly and can be used to avoid windy days and to perform flight tasks in high-risk areas, but cover small areas due to target objects and can acquire images with high resolution [15,16].

Therefore, UAV remote sensing is a flexible and efficient way of obtaining high-resolution images providing accurate information from low altitudes with less interference from clouds.

With respect to the civilian applications of UAV-based low-altitude remote sensing, since the first geomatics test was carried out by Przybilla and Wester-Ebbinghaus (1979) [17], there has been an increasing amount of interest in many fields such as forestry and agricultural resource assessment and monitoring [18,19,20]; environmental and atmospheric surveying and monitoring [21,22,23,24]; disaster monitoring and assessment [25,26,27]; three-dimensional (3D) building reconstruction and landscape mapping [13,16,28]; and 3D documentation and mapping of sites and structures of archaeological and cultural heritage [7,29,30]. Several works have been done with respect to 3D mapping and monitoring of open-pit mine areas by the use of UAVs [31,32,33]. McLeod *et al.* [34] presented a method for measuring fracture orientation in an open-pit mine by use of video images acquired from an UAV. By integrating with terrestrial laser scanning (TLS), Salvini *et al.* [31] proposed an approach for observing the structural-geological setting of a quarry wall and to identify the potentially unstable zones. The images acquired by UAV-based photogrammetric systems have the characteristics of large overlap areas, multiple viewing angles and high ground resolution [35], and, at the same time, small footprints [3], large overlap variations and large perspective distortions [16]. Therefore, with respect to the processing of UAV images, much effort has been made in the development of processing approaches and systems. Laliberte *et al.* [10] presented a workflow of using UAV for rangeland inventory, assessment and monitoring. Chiabrando *et al.* [7] described a technique for the acquisition and processing to provide digital surface models (DSMs) and large-scale maps to support archaeological studies. Zhang *et al.* [16] proposed an approach for the parallel processing of UAV images. Niethammer *et al.* [27] proposed an approach for the production of high-resolution ortho-mosaics and digital terrain models (DTMs) in landslide investigations. Harwin and Lucieer [35] assessed the accuracy of georeferenced point clouds generated via multi-view stereo (MVS) from UAV imagery. Rosnell and Honkavaara [36] proposed a method for point cloud generation by image matching using aerial image data collected by a light UAV quadrocopter system. Turner *et al.* [3] presented an approach for the geometric correction and mosaicking of UAV photography by the use of feature matching and structure-from-motion (SfM) techniques. Generally, there are two approaches for processing of UAV imagery. The first one is photogrammetric approach by the use of bundle adjustment, and the second is computer vision based approach by the use of MVS-SfM technique. The bundle adjustment usually needs more GCPs to provide a global homogeneous result over the entire area. However, it is difficult to introduce additional data to the MVS-SfM, and it is not easily accessible because the images lack texture. In the existing studies, the UAV imagery is only used to aid GCPs surveyed by GPS. TLS can capture high-accuracy dense point cloud of the object surface at high rates in near real time. Therefore, we need to integrate feature points extracted from TLS supplement as GCPs in the bundle adjustment to improve the accuracy of UAV imagery geo-positioning. Furthermore, to our best knowledge, there is not a practical framework for the integration of UAV-based photogrammetry and TLS with applications in open-pit mine areas at present.

Open-pit mine areas are usually large and are located in remote mountainous areas. As a result, traditional techniques such as global positioning systems (GPS) and electronic total station (TS), which provide the point-based observations, might sometimes have difficulty in monitoring entire large areas. The cost of these ground monitoring techniques is also rather high. Therefore, this paper is concerned with conducting a UAV campaign for the photogrammetric mapping, monitoring and assessment of open-pit mine areas and their surrounding environment. In addition, the optimization of slope angle in open-pit mine areas plays an important role in making full use of recycling resources, reducing production costs and increasing mining efficiency [37]. However, the slope zones may result in its failure. Thus, it is essential for accurate 3D mapping and monitoring of the slope zones. TLS can be used to acquire 3D point clouds of the slopes rapidly and accurately. As a result, this paper is also concerned with integrating TLS derived point clouds and UAV images for 3D mapping of the slope zones.

However, with respect to mine area inventory, assessment and monitoring by the use of UAVs, there are two critical issues that need to be further investigated. The first one is that mine areas are usually located in remote mountainous areas. Thus, it is crucial to design a specific workflow to conduct a UAV campaign for the photogrammetric mapping of open-pit mine areas and their surrounding environment. The second is that the side slope in the open-pit mine area presents irregular shapes for land surfaces and can lack texture. In order to monitor the instability of the side slopes, high-accuracy dense clouds from TLS are acquired to build an accurate geometric model of the side slope, and high-resolution UAV images are used to build a texture model of the side slope. In addition, the other areas in the open-pit mine areas, high dense clouds from UAV photogrammetry are generated to build 3D modelling of these areas with the aid of the GCPs from both GPS and TLS. For these reasons, it is necessary to integrate the high-resolution UAV images and high-quality dense point clouds from TLS to create an accurate 3D model of the side slope, and at the same time to improve the accuracy of UAV imagery geo-positioning with the aid of GCPs from both GPS and TLS to generate accurate 3D modeling of the open-pit mine areas.

The Kunyang, Jinning and Jianshan phosphate mines were founded in 1965, 1981 and 1999, respectively, and are some of the largest open-pit phosphate mines in China. Due to lengthy and extensive open-pit mining, the surrounding ecological environment has been seriously damaged, and a number of slope failures have occurred in the past decade. The water quality of Dianchi Lake near the mines has been particularly influenced. As a result, it is necessary to assess and monitor the mine areas, for the purpose of mine tailing, land reclamation, and planning the vegetation resources.

The objective of this paper is to present a framework for the integration of UAV-based photogrammetry and TLS for the 3D mapping and monitoring of open-pit mine areas and for the accurate 3D modeling of the side slopes, in southwestern China. Specifically, the focus of the paper is: (1) to demonstrate the practical workflow for UAV-based photogrammetric operations, which consists of flight planning, ground observation, image and point cloud acquisition, image and point cloud processing and integration, and 3D mapping and land-cover classification of the mine areas; (2) to investigate a method for improving the geo-positioning accuracy of UAV imagery by the use of the bundle block adjustment, with the aid of ground control points (GCPs) from a GPS survey and high-quality dense 3D point clouds from TLS; and (3) to integrate the 3D ground points from UAV images and point clouds from the TLS, aiming to provide accurate 3D mapping of the slope areas.

METHODS

The framework for the acquisition and processing of UAV images in the open-pit mine areas consists of four main parts, *i.e.*, configuration of ground network and flight route design, image acquisition and matching, accuracy improvement of UAV imagery geo-positioning and classification of land covers in open-pit mine areas from UAV ortho-images. Figure 1 shows the entire technological framework of the study. In the first part, the configuration of the ground control network includes the number and distribution of GCPs that are surveyed by the real-time kinematic global positioning system (RTK-GPS). The flight route is designed with the specific factors for UAV-based photogrammetric and aerial image orientation. In the second part, both UAV images and the position and orientation system (POS) data are acquired during autonomous flight operations. The conjugate points of a stereo pair of images are extracted by the scale-invariant feature transform (SIFT) operator [38] and the outliers are identified and eliminated by the RANdom SAmple Consensus (RANSAC) approach [39]. In the third part, with the aid of GCPs and 3D point clouds from TLS, the accuracy of the geo-positioning of UAV imagery is improved by using bundle block adjustment. In addition, accurate 3D mapping of the side slopes is generated by integration of high-resolution UAV images and high-quality 3D point clouds. In the fourth part, a multi-resolution segmentation algorithm and a bottom-up region merging technique are used for the image segmentation, and an object-based image analysis approach is used for image classification.

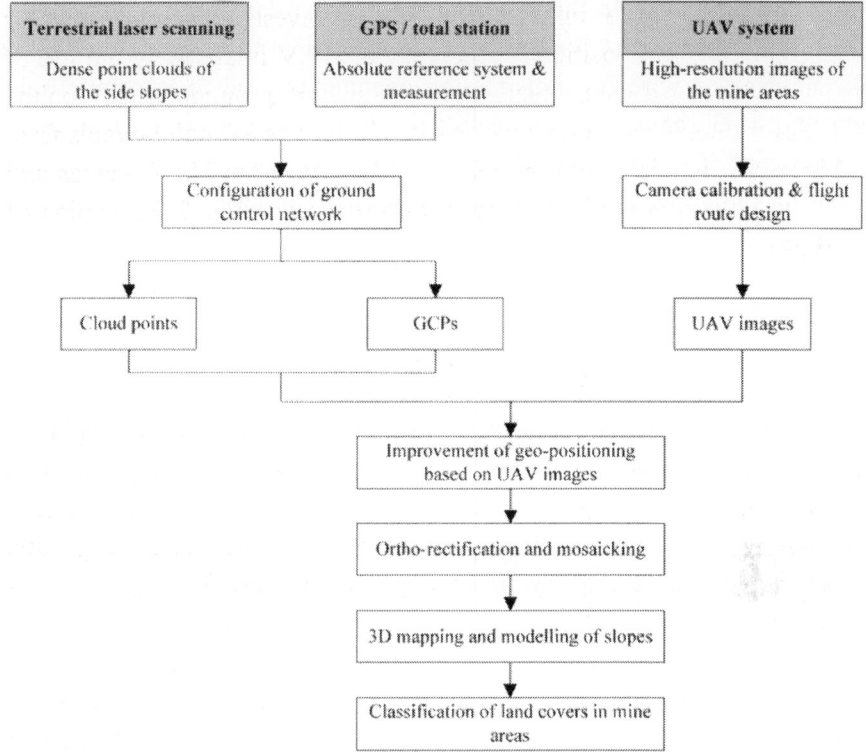

Figure 1. Entire workflow of the study.

Materials and Data Acquisition

The UAV System, Sensors and the TLS System Used in the Experiment

The UAV system used in this study is an ISAR-II UAV manufactured by the Beijing Remote Sensing & Digital Earth Information Technology Company (http://www.uavrs.com/). The system consists of two main components: aerial and ground systems (Figure 2). The aerial systems consist of the remote sensing sensor suite, the automatic control system and the UAV platform. The main functions of the aerial systems are to upload the plan routes into the controller and to monitor the state of the UAV in flight. The ground components include a route planning system, a ground control system (GCS) and a data reception system. The main functions of the ground components are to design and schedule the flight routes, to receive real-time flight attitude data and to control the flight of the UAV. In the experiment, a GPS reference station was set within the distance of 30 km from the study area to navigate the UAV. In addition, a wireless transmission channel is used to transmit flight data between the aerial systems and the ground station with a maximum distance of 30 km to monitor the UAV flight routes, to change the flight plan or switch automatically to manual flight in case of emergency or landing.

In this study, the UAV platform used was a fixed-wing miniature plane, which has a payload capacity of approximately 4 kg and a flight duration of approximately 1.4 h (limited by the 2.2-liter gas tank).Table 1 shows the technical specifications of the UAV platform. The platform is equipped with a micro-autopilot unit, enabling automated navigation and real-time monitoring of the overlooked areas. The navigation system includes a GPS/DGPS unit, a three-axis gyroscope and accelerometer, and a relative airspeed probe. The UAV platform can fly automatically, according to the predefined flight routes, based on GPS and IMU navigation. Photos are taken during the flight at a set interval of time or distance. During the flight process, the flight parameters, such as the spatial location of the UAV, the three attitude angles and flight speed, are recorded. In addition, a Dagama SiRF3 SG-959 G-mouse module was mounted on the UAV having an update rate of 1 Hz. The IMU sensors include a three-axis rate gyroscope, accelerometer and magnetometer. The accuracy of the attitude data from the IMU is rated as $\pm 2°$ for roll and pitch and $\pm 5°$ for heading. Therefore, it is necessary to improve the accuracy of geo-positioning of UAV imagery by correcting the biases in the orientation parameters with aid of GCPs obtained from GPS survey and 3D point clouds obtained from the TLS. The detailed methods will be discussed in Section 2.3.

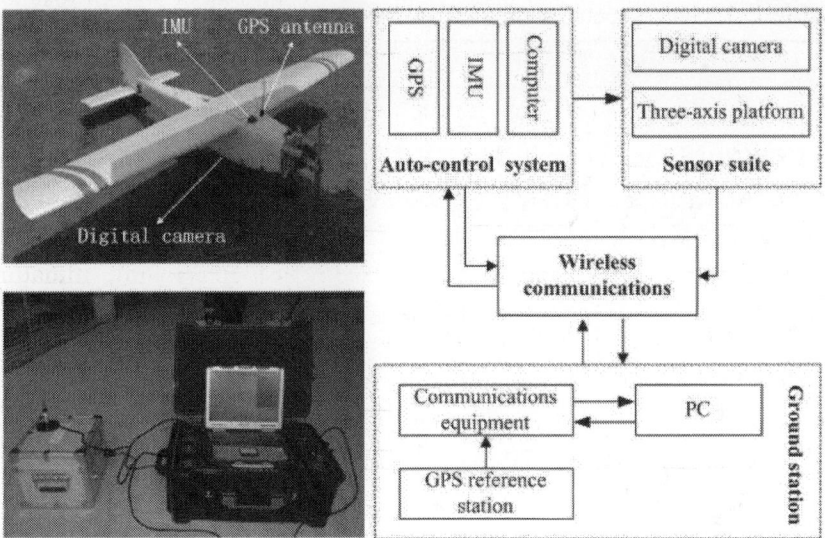

Figure 2. The UAV system used in the experiment.

In the experiment, the single lens reflex (SLR) digital camera equipped in the platform is a Canon EOS 5D Mark II camera, with a focal length of about 24 mm, image size of 5616 × 3744 pixels, and pixel size of 6.41 μm. The CCD size of the camera is 36 mm × 24 mm, with 21 million pixels, and the images acquired by the camera are true colour photo with 8-bit radiometric resolution.

A Leica HDS8800 long-range TLS was used to acquire 3D point clouds of the side slopes in the mine areas. The position and orientation of the TLS used in the experiment will be discussed in Section 2.1.2. The main parameters of the TLS are presented in Table 2.

Table 1. Technical specifications of the UAV platform used in the experiment.

Item	Value
Length (m)	1.8
Wingspan (m)	2.6
Payload (kg)	4
Take-off weight (kg)	14
Endurance (h)	1.8
Flying height (m)	300–6000
Flying speed (km/h)	80–120
Capacity	Fuel
Flight mode	Manual, semi-autonomous and autonomous
Launch	Catapult, runway
Landing	Sliding, parachute
Sensor	Digital camera, video camera

Table 2. Technical specifications of the TLS used in the experiment.

Item	Value
Range	2.5–2000 m
	1400 m to 80% albedo (rock)
	500 m to 10% albedo (coal)
Scan rate	8800 points per second
Divergence	+0.25 mrad
Range accuracy	10 mm to 200 m
	20 mm to 1000 m
Angle accuracy	±0.01°
Repeatability accuracy	8 mm

Study Area and Data Acquisition

The study area is located to the south-west of Kunming city in Yunnan province, China, and is adjacent to the southern border of Dianchi Lake (Figure 3). In the study area, there are three open-pit phosphate mines, Jianshan, Kunyang and Jinning, with areas of 7.8 km^2, 18 km^2 and 43 km^2, respectively. The elevation in the study area ranges from 1888 m to 2485 m.

Prior to the image acquisition, fight route planning is necessary for the aerial image orientation and the subsequent generation of the photogrammetric products [30]. The specific factors for photogrammetric UAV flight planning can be found in many publications [14,19,40]. In the experiment, Google Earth™ and the UAV ground station software (GSS) were used to design the flight route. After the coordinates of the flight areas were determined, the information was transferred to the GSS, and the flight lines were generated based on the desired flying heights and image overlap.

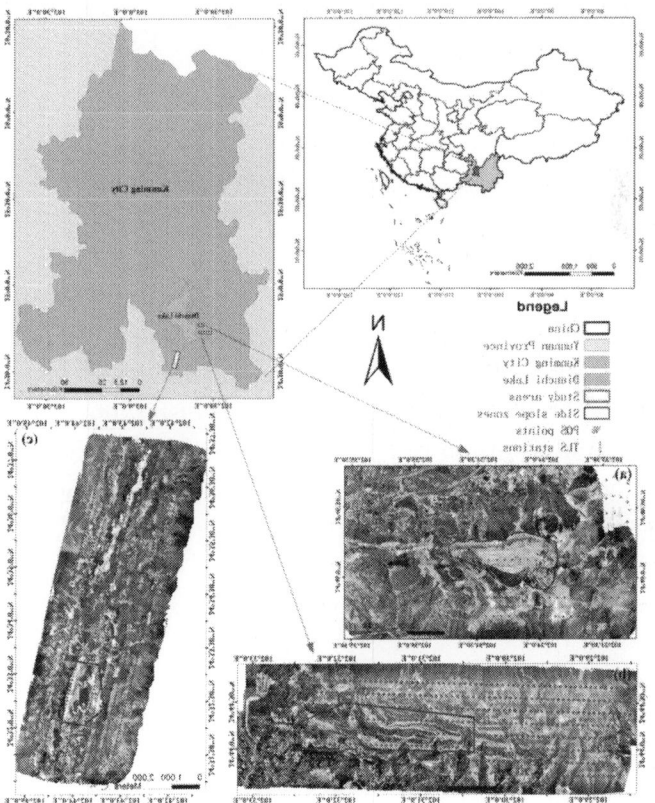

Figure 3. The study area in southwestern China. The UAV flight areas are delineated as red polygons, the side slope zones are delineated as blue polygons, the planimetric positions of the photo centres are delineated as red points, and the TLS positions are delineated as green points. (**a**) Jianshan open-pit mine area; (**b**) Kunyang open-pit mine area; and (**c**) Jinning open-pit mine area.

In the study area, during a flight time of approximately 143 minutes in May 2011, a total of 1688 UAV images were collected over the study areas, among which there were 290 images in the Jianshan area, 618 images in the Kunyang area and 780 images in the Jinning area. Table 3 shows the particular information of the data acquisition. The spatial resolution of these UAV images is approximately 15 cm, with 75% forward lap and 55% side lap. In the experiment, the UAV images were acquired during the flight operations and were stored on the camera's 16 GB memory card. For each image, a timestamp, GPS location, roll, pitch and heading were recorded by the onboard computer. The images and flight data were downloaded after landing.

Table 3. Data acquisition achieved over the study areas.

Area	Strips	Images	GCPs	Area (km^2)	Consumed Time (min)	Average Flying Height (m)
Jianshan	8	290	28	7.8	30	563.37
Kunyang	8	618	40	18	49	476.23
Jinning	8	780	29	43	64	563.37

The SLR digital camera mounted on the UAV has a non-metric lens. Prior to processing the obtained images, a digital camera distortion model was performed, which includes radial distortion (*i.e.*, k_1, k_2 and k_3), asymmetric distortion (*i.e.*, p_1, and p_2) and plane distortion (*i.e.*, α and β) in the camera calibration ([41]). Table 4 shows the result of the calibration parameters of the camera equipped on the UAV.

Table 4. Result of the calibrated parameters of the digital camera mounted on the UAV.

Item	Value	Deviation
Focal Length (mm)	24.3704	0.0001
Principal Point x_0 (mm)	0.2014	0.0001
Principal Point y_0 (mm)	0.0638	0.0001
Radial Distortion	k_1: 7.80246 e-09	2.7 e-010
	k_2: −5.20000 e-16	−3.839 e-017
Decentering Distortion	p_1: −1.10102 e-07	8.6 e-008
	p_2: 9.25639 e-08	9.3 e-009
Affinity and Nonorthogonality	α: −5.24645 e-05	–
	β: −2.78373 e-06	–

Image orthor ectification and georegistration are related to aerial triangulation based on the presence of GCPs in the project area [3,42,43,44,45]. For UAV-based photogrammetry, GCPs are focused on obvious feature points on the images, such as road intersections and house corners. However, it is sometimes difficult to find obvious features in open-pit mine areas. Therefore, in the experiment, 97 man-made markers (Figure 4), which are designed as black and flat circles (with 60 cm diameter) printed on white cloth (with the size of 2 m × 2 m), were established as GCPs and were surveyed with the RTK-GPS in the national geodetic coordinate system, with an accuracy of 2–3 cm in the horizontal direction and up to 5 cm in the vertical direction. In addition, a total of eight GCPs were simultaneously measured by both GPS-RTK and TS in Jianshan area, with the aim of assessing the accuracy of GCPs surveyed by RTK-GPS through comparing the differences between the coordinates obtained from RTK-GPS and TS. Each GCP was measured three times, and the mean value was calculated as the coordinate of each GCP. In the experiment, the root-mean-squared errors (RMSE) of the coordinate residuals between RTK-GPS and TS in the X-, Y- and Z-directions are 0.034 m, 0.0098 m and 0.0374 m, respectively. Therefore, the accuracy of the GCPs is much higher than the spatial resolution (approximately 0.15 m) of the UAV images used in the experiment.

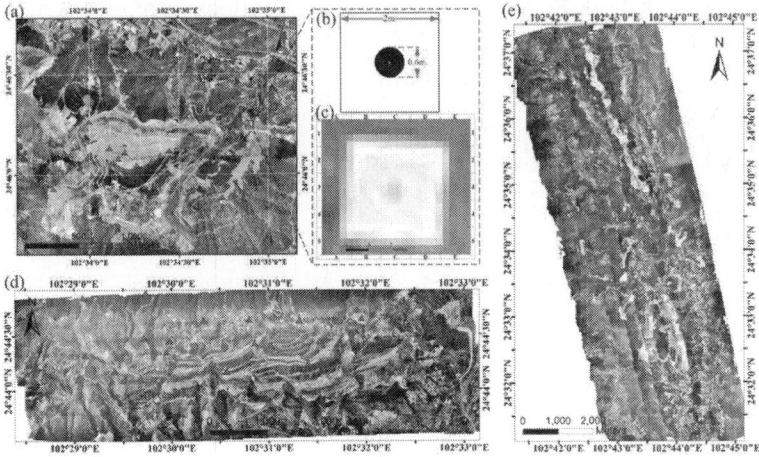

Figure 4. Distribution of GCPs in the study area. (**a**) configuration of the GCPs (delineated as red triangles) in the Jianshan area; (**b**) designed pattern of the GCPs; (**c**) example of a GCP in the UAV image; (**d**) configuration of the GCPs in the Kunyang area; (**e**) configuration of the GCPs in the Jinning area.

Moreover, the surfaces of the side slopes in the mine areas were surveyed by TLS, which is a technique of direct 3D measurement, and a total of 10,826,832 points with intensity and RGB data were obtained, having an average point space of 5 cm. In the experiment, the position and orientation of TLS were determined by the direct georeferencing approach [46], in which the known control point was measured by TS, and the maximum distance between TLS and the object surface was 700 m; thus, the accuracy of point clouds was less than 122 mm, which is consistent with the result in [46,47]. Figure 3 shows the positions of TLS used in the experiment in the three study areas. The registration of TLS point clouds was processed in Leica Cyclone, and the accuracy was 8.7 mm, which is also similar to the result in [48,49,50,51].

Image Matching and Extraction of Feature Point Cloud

Image matching is a very useful technique in any type of photogrammetry (aerial and terrestrial) and fundamental to the use of UAVs [16,52]. Traditionally, the feature-based matching operators (for example, Forstner and Harris operators) and area-based matching techniques (such as cross-correlation and least squared matching methods) are widely used in highly textured images. Furthermore, the images acquired within open-pit phosphate mines often lack texture. Therefore, it might be difficult to obtain reliable results with these traditional matching operators. In our study, the SIFT operator was used for feature extraction and matching, which has the advantages of matching image between UAV-based low-altitude image pairs with large rotation angles [16,36,52]. The SIFT features are first extracted from each image of a stereo pair, and then the two images are matched by comparing each feature in one image with all the features in the other image, and lastly conjugate points are found within the predefined searching range [16]. It is likely that there would be some mismatched points in the matching result. Therefore, in order to identify and eliminate these outliers, an iterative RANSAC approach is used to estimate the fundamental matrix (detailed in [53], p. 279) and to select the maximum number of inliers. In the experiment, the adopted approach started by using a random subsample of points to estimate the parameters that define the model, and the other points were checked with the estimated model, within a predefined threshold tolerance (*i.e.*, a threshold of three pixels in the experiment). Figure 5 shows the result of the matched conjugate points of a stereo pair of UAV images.

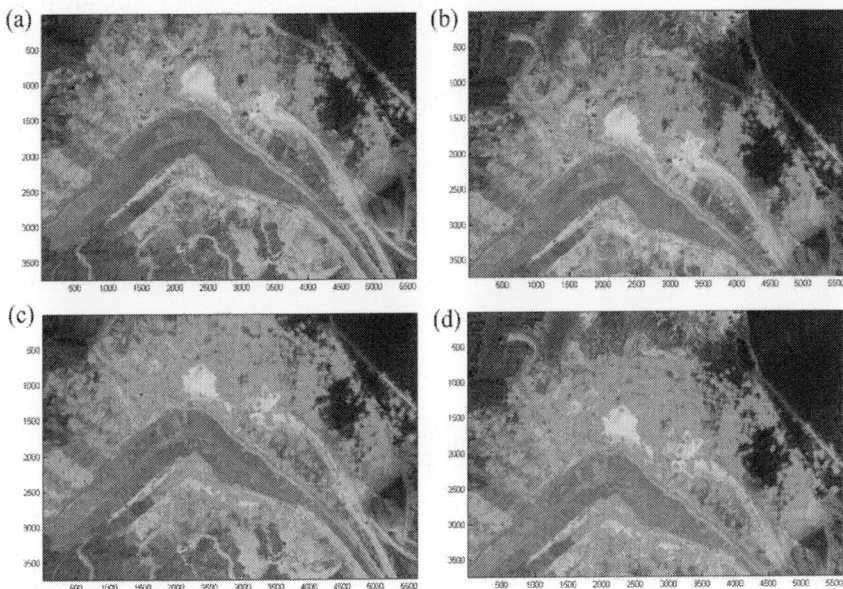

Figure 5. Result of the matched points between two UAV images in the Jianshan area (Unit: pixel). (**a,b**) matched points by the use of SIFT operator (2901 matched point pairs from a total of 57,243 point pairs with a threshold of 0.3 pixels); (**c,d**) dense matched points based on the method of the optimal exterior orientation parameters (5517 matched point pairs with a searching size of 21 pixels, correlation size of 7 pixels and least squares size of 21 pixels).

TLS and UAV photogrammetry can be complementary [54,55] because TLS can capture high-quality dense point clouds and provide accurate geometry information, and digital cameras can acquire high-resolution images and supply additional surface colour. In this paper, the integration of UAV images and 3D point clouds of TLS are studied by the use of the bundle block adjustment. The operational procedure for matching conjugate points between the TLS point clouds and the UAV images consists of four steps, as follows. (1) The TLS point clouds are first converted into an intensity image. The generated intensity image has 5500 × 1760 pixels with the cell size of 250 mm. Prior to the conversion, the singular or gross points were removed; (2) The intensity image is matched with the UAV image. During the matching process, the SIFT operator was used in the feature extraction, and a total of 141 conjugate points were extracted in the experiment; (3) The accuracy of the matched conjugate points is assessed. By the use of GCPs surveyed by RTK-GPS, the translation between the intensity image and UAV image is constructed. Thus, the image coordinates of the matched points from the UAV image is transformed into the ground coordinates, and the differences between the transformed coordinates and the coordinates of the TLS point clouds are calculated for evaluating the accuracy of the matched points; (4) The correctly matched conjugate points between the TLS point clouds and UAV images are selected based on the calculated root-mean-squared error of the residuals. In the experiment, the threshold was set as 2.35 pixels.

Geo-Positioning Accuracy Improvement and 3D Textural Modelling

The accuracy of geo-positioning of UAV images depends jointly on the camera orientation, configuration of the GCPs (*i.e.*, accuracy, density and distribution), image quality, land cover and the topographic complexity of the scene [6]. The impact of the configuration of the GCPs on geo-positioning accuracy based on high-resolution satellite imagery has been studied [35,56,57,58,59]. In aerial triangulation, bundle block adjustment provides a global homogeneous result over the entire area [60]. In order to improve the accuracy of geo-positioning based on UAV imagery in the three mine areas, the extracted feature points in the 3D point clouds obtained from the TLS are used as a supplement to the GCPs in the bundle adjustment, and three scenarios for integration of both point clouds from 3D TLS and GCPs from a GPS survey were designed as follows. (1) Scenario one involves performing the bundle adjustment only with the support of the POS; (2) Scenario two involves performing the bundle adjustment with both the POS and the GCPs surveyed by the GPS; (3) Scenario three involves performing the bundle adjustment with the POS, GCPs and 3D point clouds from the TLS. As a result, both the 3D ground coordinates of the conjugate points and the exterior orientation parameters (EOPs) of the cameras are simultaneously estimated through bundle adjustment with the GCPs from the TLS point clouds and GPS control points. Based on the improved accuracy of geo-positioning based on UAV images in aerial triangulation, a DSM over the study area was produced by the use of the Delaunay triangles which were generated from the integration of 3D ground points of the entire area obtained from the aerial triangulation and 3D point clouds of the slope zone obtained from TLS, and the ortho-images were rectified from the original UAV images with the refined camera parameters and the generated DSM. In the study, UAV images covered the entire open-pit mine areas, where trees, grasses, vegetation and exposed rock/soil were present. In addition, the slope zone was covered by the point clouds from TLS, and the area had few trees and grasses. Moreover, a filtering operation in the creation of the DSM from UAV and DSM from TLS needs to be considered. The bundle adjustment, ortho-rectification and mosaicking of the acquired images were produced by using the commercial software Leica Photogrammetry Suite (LPS, ERDAS 9.2, Leica Geosystems Geospatial Imaging, Norcross, GA, USA). In the experiment, a quadratic optimal approach of aerial triangulation was used. Firstly, the matched conjugate points obtained by the SIFT operator with high accuracy were imported in the point measurement with the required data format, and a bundle block adjustment with the initial EOPs was then conducted to estimate the optimal EOPs. Afterwards, these dense matched points were extracted based on the estimated optimal EOPs (as shown in Figure 5), and the bundle block adjustment was performed once again.

The UAV-based photogrammetric system can provide high forward lap and side lap of high-resolution images and even oblique photos. At the same time, the dense 3D point clouds from the TLS can be more appropriate to model the complicated and irregular side slopes in detail. Therefore, the integrated high-resolution images and high-quality dense point clouds can be used to generate accurately textured 3D recordings and presentations [61]. In the experiment, a 3D representation of the side slopes was built by integrating the generated DSM in the commercial software ArcScene™ (ArcGIS 9.2), based on the point clouds from the TLS and the UAV ortho-images, with the aim of providing detailed information for assessment and planning of mine areas, even for monitoring with long time data collection. The correspondence between the two datasets was established by the GCPs from the TLS point clouds and GPS control points, and the two datasets can be registered into a common reference system.

Classification of the Land Covers in the Mine Areas

With respect to classification of land covers, an object-based image analysis (OBIA) approach is suitable for high-resolution imagery, and it has been demonstrated in many studies [5,15,62,63]. The accuracy improved UAV ortho-images have a high spatial resolution, however, both spectral and radiometric resolutions are highly correlated (*i.e.*, red, green and blue bands) [64]. As a result, the intensity-hue-saturation (IHS) space, which is transformed from the RGB space, can increase the accuracy of classification [65]. In the experiment, the OBIA approach was developed in eCognition 8.64. The operational procedure for the OBIA approach consists of three steps, as follows. (1) Segmentation of the image by the use of a multi-resolution segmentation (MRS) algorithm [66]; (2) Selection of texture measures to separate the classes of interest, based on a decision tree; (3) Determination of the class separability and the accuracy of the classification of the land covers. In the MRS algorithm, a bottom-up region merging technique was used to combine the smaller segments into larger ones based on the relative homogeneity criteria such as scale, color and shape. Figure 6 shows the result of image segmentation at different scales. In the figure, the MRS is based on a local homogeneity criteria to describe the similarity of the adjacent image objects.

The UAV images, which acquired in the open-pit phosphate mines that are located in the mountain areas, showed more trees and grasses. In the experiment, a combination of rule-based and nearest-neighbour classification methods was used to distinguish the classes. Three types, which are the vegetation (Woodland, Grassland and Arable land), exposed rock/soil (Building, Road, Active mine, Reclaimed mine and Rural area) and water, were first classified using the rule-based classification with the defined thresholds of intensity, and the detailed classification at the species level was conducted using the nearest-neighbour classification method. With respect to the vegetation, an experiential vegetation index (VI') was first calculated by $VI'=(2 \times G' - R' - B') - (1.4 \times R' - G')$, where $G' = G/(R+G+B)$, $R' = R/(R+G+B)$ and $B' = B/(R+G+B)$, respectively. If the value of the calculated VI' is greater than the threshold of -0.1, then the vegetation type can be determined. With respect to the water, a grey level co-occurrence matrix (GLCM) homogeneity [15,67] was further calculated. If the value of GLCM is greater than 0.37 and the value of the calculated *VI'* is not between −0.28 and −0.15, then the water type can be determined accordingly. The accuracy of the classification was assessed by error matrices, and thus, by calculating the overall user's (errors of omission) and producer's (error of commission) accuracies, overall accuracy, as well as Kappa statistics [68]. In the accuracy assessment of the classification of the land covers, at least 10 point samples for each class were manually labeled as the ground truths (details can be seen in [69]). In addition, more than 100 point samples were selected in the active areas, reclaimed mines, woodland and grassland, and these samples were distributed evenly across the entire study area.

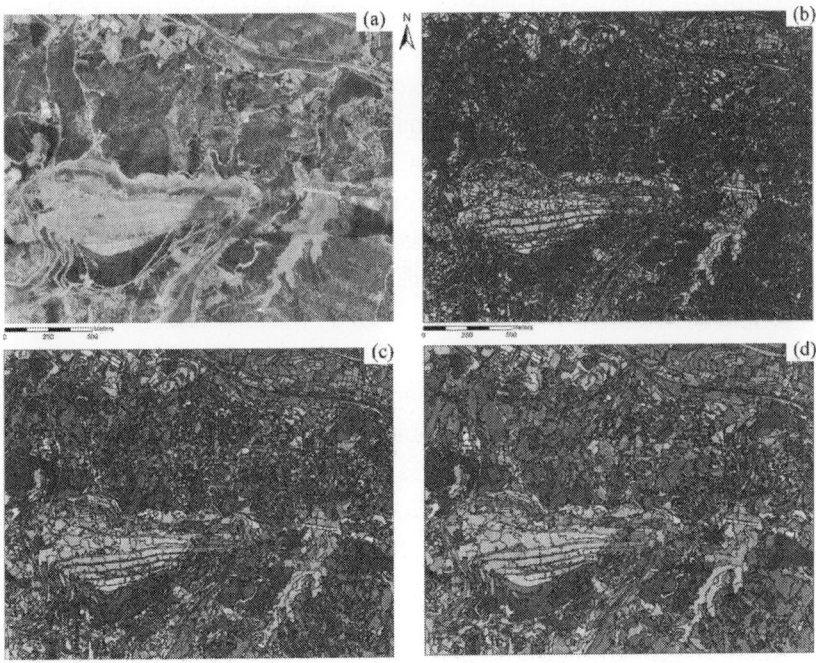

Figure 6. Comparison of image segmentation at the different levels used in the object-based classification. (**a**) UAV mosaic image from Jianshan; (**b**) Image segmentation with a MRS scale of 200; (**c**) Image segmentation with a MRS scale of 400; (**d**) Image segmentation with a MRS scale of 600.

RESULTS AND DISCUSSION

Result of the Geo-Positioning and Accuracy Assessment

With respect to the integration of UAV-based photogrammetry and TLS, Table 5 shows a comparison of the geo-positioning accuracy based on UAV images in the three scenarios, as discussed in Section 2.3, of the entire Jianshan area. In the test, 10 point clouds from the TLS (detailed see Section 2.2) and 28 GCPs surveyed by the GPS, in the Jianshan area, were used in the bundle block adjustment (in which nine GCPs were used as checkpoints to assess the accuracy of the geo-positioning). There were six point clouds from the TLS and 32 GCPs (in which eight GCPs were used as checkpoints) surveyed by the GPS in the Kunyang area. While in the Jinning area, only 21 GCPs (in which eight GCPs were used as checkpoints) surveyed by the GPS were used in the bundle block adjustment. From the table, we can see that: (1) the geo-positioning accuracies in both scenario two and three are much better than that in scenario one. This result indicates that in the steep slope zone area, the accuracy of geo-positioning based on UAV images only with the support of the POS in scenario one is relatively low. However, with the aid of more GCPs from the GPS or TLS, the geo-positioning accuracy can be significantly improved; (2) Scenario three achieves the best accuracy of the geo-positioning. The reason for this is that some accurate 3D point clouds obtained from the TLS are used as a supplement to the GCPs in the bundle adjustment. In general, the decimeter-level accuracy achievable from UAV images meets the demand for the subsequent application of images in mine areas; (3) The accuracy in Z-direction with GCP demonstrated more improvement; it may be there were more errors in the initial value of POS in Z-direction than in the planimetric. However, there remain more errors in the Z-direction, which may be influenced by the steep slope, and the images may have larger perspective distortions.

Table 5. Results of the geo-positioning accuracy with bundle block adjustment.

Area	Scenario	Maximum Error			Mean Error			Root-Mean-Squared Error		
		X/m	Y/m	Z/m	$|X|$/m	$|Y|$/m	$|Z|$/m	X/m	Y/m	Z/m
Jianshan	Scenario one	−4.4216	2.1685	11.6474	3.8868	1.7411	8.0782	3.9091	1.7597	8.5113
	Scenario two	−0.2586	0.1391	1.4753	0.1116	0.0842	0.4850	0.1407	0.0949	0.6380
	Scenario three	−0.2477	−0.1488	1.6074	0.0974	0.0827	0.4183	0.1253	0.0933	0.6210
Kunyang	Scenario one	−4.2221	−7.0257	−16.7015	1.4489	4.4386	6.6686	1.8331	4.6088	8.0069
	Scenario two	−0.2529	0.2839	−1.4345	0.0990	0.1248	0.4875	0.1414	0.1447	0.7766
	Scenario three	−0.2250	0.2284	−2.4785	0.1095	0.1262	0.5402	0.1378	0.1416	0.7080
Jinning	Scenario one	2.2202	2.3067	−5.0525	0.4131	1.2843	1.4831	0.7088	1.4342	1.9025
	Scenario two	−0.3473	−0.8749	1.1805	0.1646	0.3344	0.6137	0.2153	0.4496	0.7813

In addition, to compare the contribution of the 3D point clouds from the TLS in the accuracy improvement in the bundle block adjustment, three tests were further conducted in the side slope zone of the Jianshan area, where there are more 3D point clouds from the TLS that are used as control points in the bundle adjustment. (1) Test one involves performing the bundle adjustment only with the support of the POS; (2) Test two involves performing the bundle adjustment with both the POS and 3D point clouds from the TLS; (3) Test three involves performing the bundle adjustment with the POS and the GCPs surveyed by the GPS. Table 6 shows the results of the geo-positioning accuracy in the three scenarios. In the test, a total of 19 UAV images, 10 point clouds from the TLS and 16 GCPs surveyed by the GPS were used in the bundle block adjustment. Among which, six GCPs were used as checkpoints, and the accuracy of the geo-positioning was evaluated by the use of these check points through comparing the difference between their coordinates obtained from the bundle adjustment and those of surveyed by RTK-GPS. From the table, we can see that the geo-positioning accuracies in both Scenarios two and three are much better than that in Scenario one. This result shows that the introduction of 3D point clouds from the TLS as GCPs can improve the accuracy of geo-positioning based on UAV images. At the same time, the accuracy of geo-positioning based on 3D point clouds from the TLS is closer to that based on GCPs from the GPS survey. This result shows the potential of using 3D point clouds from the TLS as GCPs in bundle adjustment, in the case where it is difficult to conduct a GPS survey in mountainous and remote mine areas, such as the slope body with steeper angle and other high-risk areas.

Table 6. Results of the geo-positioning accuracy with bundle block adjustment in the side slope zone of the Jianshan area.

Test	Maximum Error			Mean Error			Root-Mean-Squared Error		
	X/m	Y/m	Z/m	$|X|$/m	$|Y|$/m	$|Z|$/m	X/m	Y/m	Z/m
Test one	0.563	0.741	1.760	0.4108	0.3647	1.4427	0.4234	0.4178	1.6589
Test two	0.227	0.376	1.507	0.1540	0.1953	1.3078	0.1634	0.2260	1.5662
Test three	0.163	0.372	1.472	0.1048	0.1448	1.0603	0.1108	0.1726	1.3209

Results of the Digital Photogrammetric Products and the Accuracy Assessment

Results of the DSMs and DOMs for the Mine Areas

Once aerial triangulation is completed, a DSM can be generated from the 3D ground points by the methods of multi-image dense matching and forward intersection. After that, a digital orthophoto map (DOM) can be generated by ortho rectification and mosaicking, based on the generated DSM and the known camera parameters. Figure 7 shows the results of the generated DSM (left) and ortho-image (right) in the Jianshan area, respectively.

In the experiment, the accuracies of the generated DSMs and DOMs were assessed. The coordinates of the checkpoints were first manually measured from the generated DOMs and DSMs, and they were then compared with the corresponding ground coordinates of these checkpoints surveyed from the GPS. Thus, the root-mean-squared error was further calculated based on the difference between the measured coordinates from the generated DOMs and DSMs and the surveyed coordinates from the GPS. Table 7shows the results of the accuracy assessment of the generated DSMs and DOMs in the study areas. In the table, the accuracies of the generated DOMs in the X- and Y-directions are calculated, as well as the accuracies of the generated DSMs in the Z-direction. From the table, we can see that: (1) In all three areas, the accuracy in the X-direction is higher than those in both the Y- and Z-directions. In addition, the accuracy of the generated DSMs is two to three times higher than that of the generated DOMs; (2) In the horizontal direction, the accuracy of the generated DOM in the Jianshan area is the best and is close to the resolution of the UAV imagery. The reason for this result might be due to a better configuration of GCPs and a higher density of GCPs in this area; (3) In the Z-direction, the error in the value of Table 7 is higher than that of in Table 5, and the accuracy is lower than those of in both X- and Y-directions, in Jianshan area. The reason for this result may be some biases due to ortho-rectification, smoothing and the resampling introduced in the process of generating DSMs and DOMs, the accuracy of geo-positioning based on UAV images is related to topographic terrain. Because this test was conducted in the steep slope zone area, the accuracy in the Z-direction is lower than those of in both X- and Y-directions. For another, apart from active mines, the vegetation areas (occupied by woodlands) would have some impacts on the image matching and the creation of DSMs. However, the image matching algorithms in vegetated area were not addressed.

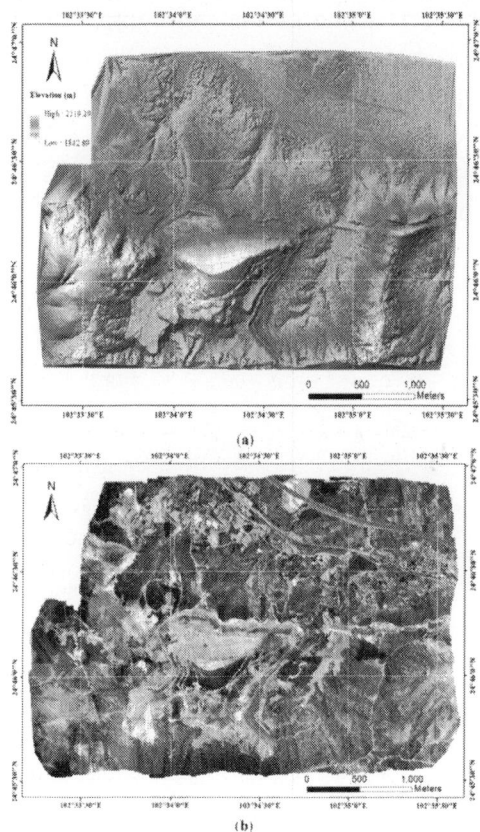

Figure 7. Result of the generated. (a) Hill-shaded DSM (with a cell size of 2 m) and (b) the generated DOM (with a spatial resolution of 0.15 m) in Jianshan area.

Table 7. Results of the accuracy assessment of the generated DSMs and DOMs in respect to GCPs from GPS in the study areas.

Area	Maximum Error			Mean Error			Root-Mean-Squared Error		
	X/m	Y/m	Z/m	$\|X\|$/m	$\|Y\|$/m	$\|Z\|$/m	X/m	Y/m	Z/m
Jianshan	0.2923	−0.2626	1.8293	0.1055	0.0773	0.5150	0.1334	0.1091	0.7245
Kunyang	0.4344	−0.2529	−1.6676	0.1152	0.1364	0.5510	0.1484	0.1545	0.7112
Jinning	−0.8437	−1.0807	1.4439	0.2382	0.4277	0.7142	0.3034	0.5104	0.8161

Results of the 3D Texture Models of the Side Slopes

The 3D texture model is one of the most important photogrammetric products, and it provides useful information for the monitoring, assessment and planning of mine areas. Furthermore, 3D models with fine textures were generated by integrating the UAV images and the point clouds of the TLS. Figure 8 shows the results of the 3D texture models of the side slopes in the study areas.

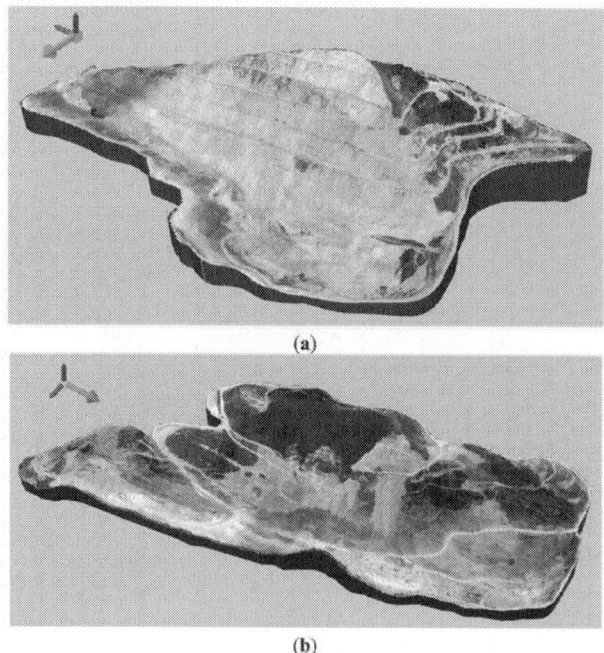

(a)

(b)

Figure 8. Result of the 3D texture model of the side slopes in the study area. (a,b) sided views from the top-left viewpoint with approximately 45° of the side slopes in Jianshan and Jinning areas, respectively.

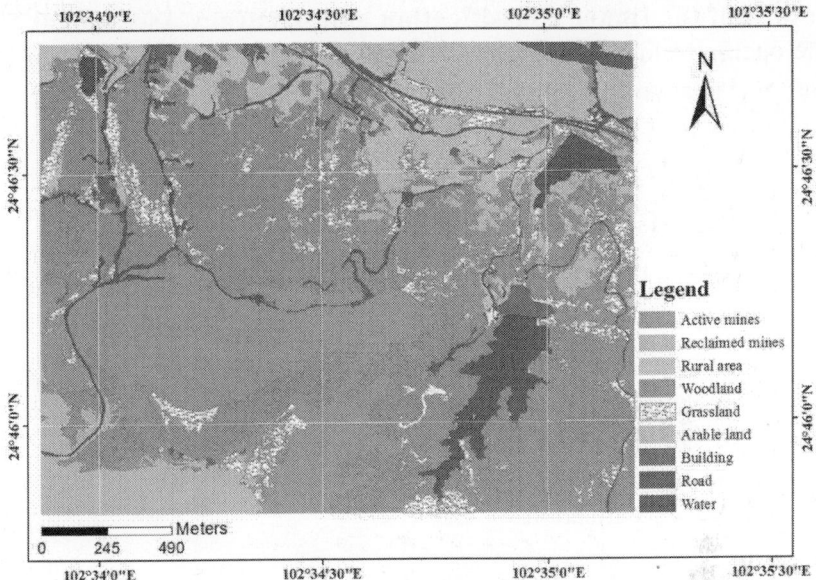

Figure 9. Result of the classification of land covers in the Jianshan area.

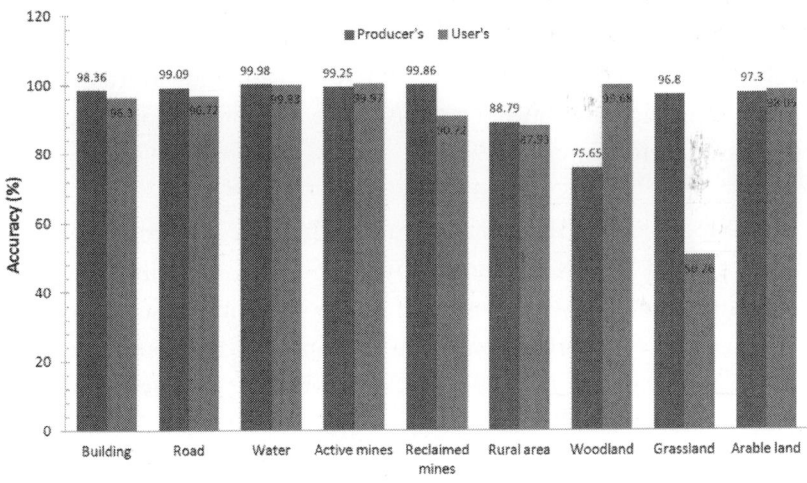

Figure 10. Accuracy assessment of the classification of land covers in the Jianshan area.

Results of the Image Classification and Accuracy Assessment
The open-pit mine area of Jianshan will be the first area designated for land reclamation. Therefore, the paper is focusing on this area for monitoring and mapping its environment and vegetation changes. Additionally, the OBIA approach was tested in this area. Figure 9 shows the result of the classification of land covers based on UAV images in the Jianshan area. Figure 10 shows the accuracy assessment of classification of the land covers in the Jianshan area. The producer accuracy for woodland is 75.65%, which is lower than any other classes. The user accuracy for grasslands is 50.26%, which is significantly lower than any other classes. This may be the grasslands confused with the woodland. The overall accuracy of the classification was 90.67% and the Kappa coefficient was 0.89, which is calculated with the confusion matrix of the classification.

Discussion
This study explored a photogrammetric approach for the joint use of UAV imagery and TLS point cloud by introducing additional input data from TLS in the bundle adjustment of UAV imagery, with the purpose of improving the geo-positioning accuracy of UAV images, particularly in the case where it is difficult to conduct a GPS survey in mountainous and high-risk areas. Based on the result of the aforementioned experiment in the open-pit mine areas, there are four issues that need to be discussed as follows.

1)   By integration of 3D points extracted from the point clouds of TLS and ground points surveyed by GPS as GCPs in the bundle adjustment of UAV imagery, the geo-positioning accuracy of UAV images can be improved in open-pit mine areas. In our experiment, with the aid of POS equipped in UAV, the geo-positioning accuracies of UAV imagery are 4.29 m and 8.51 m in the planimetric and height directions, which are corresponding to 28.6 pixels and 56.7 pixels at the spatial resolution of 0.15 m of UAV imagery. Further, with the aid of POS, 19 GCPs from GPS survey and 10 3D points extracted from the point clouds of TLS, the accuracies can be improved to be 0.16 m and 0.62 m in the planimetric and height directions, which are corresponding to 1.07 pixels and 4.13 pixels with respect to the spatial resolution of 0.15

m of UAV imagery. In the existing studies on the geo-positioning accuracy of UAV imagery, by the use of a total of 33 GCPs in urban areas, Zhang *et al.* [16] achieved an accuracy of 0.02 m in planimetric direction and an accuracy of 0.03 m in vertical direction, which are corresponding to 0.4 pixels and 0.6 pixels compared with the 0.05 m ground sampling distance (GSD) of UAV imagery. Another result showed that the RMSEs of both planimetric position and height were better than 0.2 m, which are corresponding to 1 pixels with the 0.2 m GSD. In rangeland areas, Laliberte *et al.* [10] obtained an accuracy of 1.5 to 2 pixels in both planimetric and height directions. Laliberte *et al.* [65] showed the results of geometric accuracy in a relatively flat area with the RMSE of 0.65 m (corresponding to 10.8 pixels) and in the greater elevation difference area with the RMSE of 1.14 m (corresponding to 19 pixels), with the 6 cm GSD. In archaeological areas, Chiabrando *et al.* [7] achieved the standard deviations of 0.04 m, −0.034 m and 0.038 m in the X-, Y- and Z-directions, respectively, which are corresponding to 1 pixels, 0.85 pixels and 0.95 pixels, respectively, compared with the 0.04 m GSD. Therefore, from results of the geo-positioning accuracy based on our proposed approach and the existing ones, we can see that (1) the planimetric accuracy obtained in mine areas is higher than that obtained in rangeland areas, while the height accuracy obtained in mine areas is lower than that obtained in rangeland areas. The reason for this result might be that by introducing additional GCPs from TLS cloud points in our proposed approach, the number of GCPs used in the bundle adjustment of UAV imagery in mine areas is more than that in rangeland areas. However, due to the elevation change in open-pit mine areas, the vertical accuracy in mine areas is lower than that in rangeland areas with the flat topography. (2) Both planimetric and height accuracies obtained in mine areas are lower than those obtained in urban areas. The reason responsible for this result might be that there are more obvious ground features in urban areas than in open-pit mine areas, the topography is more flat in urban areas than in open-pit mine area, and it is easier to obtain more GCPs in urban areas than in open-pit mine areas. At the same time, the accuracy of geo-positioning achievable from UAV images based on our proposed approach meets the demand for the subsequent applications in open-pit mine areas.

2) The contribution of 3D point clouds from the TLS on the improvement of geo-positioning accuracy of UAV imagery shows the same level as that of GCPs surveyed by GPS. In our experiment in the side slope zone of the Jianshan area, with the aid of the POS and the GCPs surveyed by the GPS, the geo-positioning accuracies based on UAV imagery achieve 0.11 m, 0.17 m and 1.32 m in the X-, Y- and Z-directions, respectively. At the same time, the geo-positioning accuracies based on UAV imagery achieve 0.16 m, 0.23 m and 1.57 m in the X-, Y- and Z-directions, respectively, with the aid of the POS and 3D point clouds from the TLS. Therefore, the result shows that the accuracy of geo-positioning based on 3D point clouds from the TLS is closer to that based on GCPs from GPS survey. Further, this result shows the potential of using 3D point clouds from the TLS as GCPs in bundle adjustment, in the case where it is difficult to conduct a GPS survey in mountainous and remote mine areas. To the best of our knowledge, there is no similar report on the contribution of 3D point clouds obtained from the TLS as a supplement to the GCPs on the improvement of geo-positioning accuracy of UAV imagery.

3) By the use of the improved geo-positioning accuracy of UAV images, the accuracies achieved of the generated DOMs and DSMs may be about 0.13 m, 0.11 m and 0.72 m in the X-, Y- and Z-directions, with respect to the image resolution of approximately 0.15 m/pixel. In addition, an overall accuracy of the classification of the land covers is 90.67%, by the use of an object-based image analysis approach in mine areas. Laliberte et al. [10] achieved an accuracy of 1.5 to 2 m for the image mosaics with respect to the spatial resolution of 15 cm of the orthophoto, and overall classification accuracies for the two image mosaics were 83% and 88% in the rangeland area. By the use of UAV imagery with a spatial resolution of 21.8 cm, Dunford et al. [70] obtained an overall accuracy of 63% for classification of the vegetation units in species mapping in Mediterranean riparian forest. In addition, with respect to the accuracy of the classification of land covers in the open-pit mine area, the vegetation and exposed rock/soil can be classified

with high accuracy by the use of the rule-based classification method. However, the classification accuracy of grasslands and woodland is significantly lower than those of the other land cover types due to the confusion of grasslands with the woodland in the mine areas.

4) Our study demonstrates a practical framework for the integration of UAV-based photogrammetry and TLS with application to the open-pit mine areas, which includes UAV image and TLS cloud point acquisition, image and cloud point processing and integration, object-oriented classification and three-dimensional (3D) mapping and monitoring of open-pit mine areas. Actually, the UAV-based photogrammetry has been widely employed in many fields, and some work has been reported in mine areas [31,32,33]. From the results of our experiments, we can see that UAVs, which need the permission of flight regulation, can be deployed quickly and repeatedly, flying with low altitude with less interference of clouds, and costing less than satellites and manned aircrafts. The novelty of the proposed framework is the joint use of UAV-based photogrammetry and TLS by introducing additional input data from TLS as GCPs can provide more accurate and detailed UAV images for monitoring of the mine areas. However, there are several limitations with respect to the proposed approach. The first one is that it is sometimes difficult to find ground characteristic features in the open-pit mine areas. As a result, some man-made markers need to be established as GCPs. The second one is that due to battery duration, UAV campaign needs more flight routes to cover a larger area. In addition, due to the limited payload capacity, only light and limit sensors can be equipped in UAVs. The third one is that due to the larger amount of UAV imagery and TLS point clouds, high performance parallel processing computation needs to be further developed.

CONCLUSIONS

This paper investigates a practical framework for the integration of unmanned aerial vehicle (UAV) imagery and terrestrial laser scanning (TLS) for the three-dimensional (3D) mapping and monitoring of open-pit mine areas, which includes flight planning, image acquisition, image processing, TLS point clouds and UAV images integration, and classification of land covers. The performance of the proposed approach is demonstrated in three open-pit phosphate mines in Yunnan province, China. In the proposed framework, (1) in order to extract the conjugate points of the stereo pair of UAV images and those points between TLS point clouds and UAV images, the feature points were first extracted by the scale-invariant feature transform (SIFT) operator and the outliers were identified and thus eliminated by the RANdom SAmple Consensus (RANSAC) approach; (2) In order to improve the accuracy of geo-positioning based on UAV imagery, the feature points extracted from TLS used as a supplement to ground control points (GCPs) and GCPs surveyed from global positioning systems (GPS) were integrated in the bundle adjustment, and three scenarios were designed and compared; (3) In order to monitor and map the mine areas for land reclamation, an object-based image analysis approach was used for the classification of the accuracy improved UAV ortho-image. The results showed that:

1) In the case of performing the bundle adjustment only with the support of the position and orientation system (POS) in the Jianshan area, the geo-positioning accuracies achieved based on UAV imagery are 3.91 m, 1.76 m and 8.51 m in the X-, Y- and Z-directions, respectively. In the case of performing the bundle adjustment with both the POS and the GCPs surveyed by the GPS, the geo-positioning accuracies achieved based on UAV imagery are 0.14 m, 0.09 m and 0.64 m in the X-, Y- and Z-directions, respectively. In the case of performing the bundle adjustment with POS, GCPs and 3D point clouds from the TLS, the geo-positioning accuracies based on UAV imagery achieved are 0.13 m, 0.09 m and 0.62 m in the X-, Y- and Z-directions, respectively.

2) For the test of the contribution of the 3D point clouds from the TLS in the accuracy improvement in the bundle block adjustment in the side slope zone of the Jianshan area, the geo-positioning accuracies achieved based on UAV imagery are 0.16 m, 0.23 m and 1.57 m in the X-, Y- and Z-directions, respectively, with the aid of the POS and 3D point clouds from the TLS. In addition, with the aid of the POS and the GCPs surveyed by the GPS, the geo-positioning accuracies

achieved based on UAV imagery are 0.11 m, 0.17 m and 1.32 m in the X-, Y- and Z-directions, respectively.

3) With the use of the improved geo-positioning based on UAV images in the bundle adjustment, an accuracy of the decimeter-level was achieved for the generated digital surface models (DSMs) and digital orthophoto maps (DOMs) in the study areas, and the overall accuracy of the classification of the land covers was 90.67%, based on an object-based image analysis approach in the mine areas.

From the aforementioned experimental results with UAVs in the open-pit mines, we can discuss the following: (1) The proposed UAV-based photogtammetric system shows a flexible and efficient way of obtaining high-resolution images and of generating high accuracy of DSMs and DOMs. At the same time, the proposed UAV-based photogtammetric system costs less than those based on the satellites and manned aircrafts; (2) The experimental result shows the potential of using 3D point clouds from the TLS as GCPs in bundle adjustment for the improvement of geo-positioning accuracy based on UAV imagery, particularly in the case where it is difficult to conduct a GPS survey in mountainous and high-risk areas; (3) The proposed framework for the joint use of UAV-based photogrammetry and TLS, which is a photogrammetric approach with additional input data from TLS, can provide detailed information for monitoring, assessment, and planning of the mine areas with a high accuracy and frequent data acquisition. However, it is sometimes difficult to find obvious features in open-pit mine areas. As a result, some man-made markers need to be established as GCPs. In addition, some approaches for parallel processing of the large amount of TLS point clouds and UAV images need to be further developed in future work.

ACKNOWLEDGMENTS

The work described in this paper was substantially supported by the National Natural Science Foundation of China (Project No. 41325005, 41171352 and 41401531), the China Special Funds for Meteorological and Surveying, Mapping and Geoinformation Research in the Public Interest (Project No. HY14122136 and GYHY201306055), the National

High-tech Research and Development Program (Project No. 2012AA121302), and the Fundamental Research Funds for the Central Universities.

AUTHOR CONTRIBUTIONS

The theoretical framework for the integration of UAV-based photogrammetry and terrestrial laser scanning for mapping and monitoring open-pit mine areas were presented by Xiaohua Tong, Xiangfeng Liu, Peng Chen and Shijie Liu. The approach for improving the geo-positioning accuracy was proposed by Xiaohua Tong, Xiangfeng Liu, Peng Chen and Shijie Liu. The photogrammetric processing of UAV imagery was conducted by Kuifeng Luan, Lingyun Li and Shuang Liu. The approach for integration of UAV imagery and TLS point clouds was studied by Xianglei Liu, Huan Xie, Yanmin Jin and Zhonghua Hong. The experiments were performed by Xiangfeng Liu, Peng Chen, Shijie Liu, Kuifeng Luan, Lingyun Li, Shuang Liu and Xianglei Liu. All authors have contributed significantly and have participated sufficiently to take the responsibility for this research. Moreover, all authors are in agreement with the submitted and accepted versions of the publication.

REFERENCES

1. American Institute of Aeronautics and Astronautics (AIAA). *Committee of Standards "Terminology for Unmanned Aerial Vehicles and Remotely Operated Aircraft"*; AIAA: Reston, VA, USA, 2004.
2. Zhou, G.Q. Geo-referencing of video flow from small low-cost civilian UAV. *IEEE Trans. Autom. Sci. Eng.* 2010, *7*, 156–166.
3. Turner, D.; Lucieer, A.; Watson, C. An automated technique for generating georectified mosaics from ultra-high resolution unmanned aerial vehicle (UAV) imagery, based on Structure from Motion (SfM) point clouds. *Remote Sens.* 2012, *4*, 1392–1410.
4. Blyenburgh, P.V. UAVs: An overview. *Air Space Eur.* 1999, *1*, 43–47.
5. Rango, A.; Laliberte, A.; Steele, C.; Herrick, J.E.; Bestelmeyer, B.; Schmugge, T.; Roanhorse, A.; Jenkins, V. Using unmanned aerial

vehicles for rangelands: Current applications and future potentials. *Environ. Pract.* 2006, *8*, 159–168.

6. Vericat, D.; Brasington, J.; Wheaton, J.; Cowie, M. Accuracy assessment of aerial photographs acquired using lighter-than-air blimps: Low-cost tools for mapping river corridors. *River. Res. Applic.* 2009, *25*, 985–1000.

7. Chiabrando, F.; Nex, F.; Piatti, D.; Rinaudo, F. UAV and RPV systems for photogrammetric surveys in archaelogical areas: Two tests in the Piedmont region (Italy). *J. Archaeol. Sci.* 2011, *38*, 697–710.

8. Watts, A.C.; Ambrosia, V.G.; Hinkley, E.A. Unmanned aircraft systems in remote sensing and scientific research: Classification and considerations of use. *Remote Sens.* 2012, *4*, 1671–1692.

9. Nagai, M.; Chen, T.; Shibasaki, R.; Kumugai, H.; Ahmed, A. UAV-borne 3-D mapping system by multisensory integration. *IEEE Trans. Geosci. Remote Sens.* 2009, *47*, 701–708.

10. Laliberte, A.S.; Herrick, J.E.; Rango, A.; Winters, C. Acquisition, orthorectification, and object-based classification of unmanned aerial vehicle (UAV) imagery for rangeland monitoring.*Photogramm. Eng. Remote Sens.* 2010, *76*, 661–672.

11. Rango, A.; Laliberte, A.S. Impact of flight regulations on effective use of unmanned aircraft systems for natural resources applications. *J. Appl. Remote Sens.* 2010, *4*, 043539.

12. Peterson, D.L.; Brass, J.A.; Smith, W.H.; Langford, G.; Wegener, S.; Dunagan, S.; Hammer, P.; Snook, K. Platform options of free-flying satellites, UAVs or the International Space Station for remote sensing assessment of the littoral zone. *Int. J. Remote Sens.* 2003, *24*, 2785–2804.

13. Wang, J.Z.; Li, C.M. Acquisition of UAV images and the application in 3D city modeling. *Proc. SPIE* 2008.

14. Eisenbeiss, H. UAV Photogrammetry. Ph.D. Thesis, University of Technology Dresden, Dresden, Germany, 2009.

15. Laliberte, A.S.; Rango, A. Texture and scale in object-based analysis of subdecimeter resolution unmanned aerial vehicle (UAV) imagery. *IEEE Trans. Geosci. Remote Sens.* 2009, *47*, 761–770.

16. Zhang, Y.J.; Xiong, J.X.; Hao, L.J. Photogrammetric processing of low-altitude images acquired by unpiloted aerial vehicles. *Photogramm. Rec.* 2011, *26*, 190–211.

17. Przybilla, H.-J.; Wester-Ebbinghaus, W. Bildflug Mit Ferngelenktem Kleinflugzeug. In*Bildmessung und Luftbildwesen*; Zeitschrift fuer Photogrammetrie und Fernerkundung, Herbert Wichman Verlag: Karlsruhe, Germany, 1979; Volume 47, pp. 137–142.

18. Herwitz, S.R.; Johnson, L.F.; Dunagan, S.E.; Higgins, R.G.; Sullivan, D.V.; Zheng, J.; Lobitz, B.M.; Leung, J.G.; Gallmeyer, B.A.; Aoyagi,

M.; *et al.* Imaging from an unmanned aerial vehicle: Agricultural surveillance and decision support. *Comput. Electron. Agric.* 2004, *44*, 49–61.

19. Xiang, H.; Tian, L. Development of a low-cost agricultural remote sensing system based on an autonomous unmanned aerial vehicle (UAV). *Biosyst. Eng.* 2011, *108*, 174–190.

20. Wallace, L.; Lucieer, A.; Watson, C.; Turner, D. Development of a UAV-LiDAR system with application to forest inventory. *Remote Sens.* 2012, *4*, 1519–1543.

21. Watai, T.; Machida, T.; Ishizaki, N.; Inoue, G. A lightweight observation system for atmospheric carbon dioxide concentration using a small unmanned aerial vehicle. *J. Atmos. Ocean. Technol.* 2006, *23*, 700–710.

22. McGonigle, A.J.S.; Aiuppa, A.; Giudice, G.; Tamburello, G.; Hodson, A.J.; Gurrieri, S. Unmanned aerial vehicle measurements of volcanic carbon dioxide fluxes. *Geophys. Res. Lett.* 2008, *35*, L06303.

23. Hardin, P.J.; Jensen, R.R. Small-scale unmanned aerial vehicles in environmental remote sensing: Challenges and opportunities. *GISci. Remote Sens.* 2011, *48*, 99–111.

24. Khan, A.; Schaefer, D.; Tao, L.; Miller, D.J.; Sun, K.; Zondlo, M.A.; Harrison, W.A.; Roscoe, B.; Lary, D.J. Low power greenhouse gas sensors for unmanned aerial vehicles. *Remote Sens.* 2012, *4*, 1355–1368.

25. Zhang, Z.X.; Zhang, Y.J.; Ke, T.; Guo, D.H. Photogrammetry for first response in Wenchuan earthquake. *Photogramm. Eng. Remote Sens.* 2009, *75*, 510–513.

26. Zhou, G.Q. Near real-time orthorectification and mosaic of small UAV video flow for time-critical event response. *IEEE Trans. Geosci. Remote Sens.* 2009, *47*, 739–747.

27. Niethammer, U.; James, M.R.; Rothmund, S.; Travelletti, J.; Joswig, M. UAV-based remote sensing of the Super-Sauze landslide: Evaluation and results. *Eng. Geol.* 2012, *128*, 2–11.

28. Marzolff, I.; Poesen, J. The potential of 3D gully monitoring with GIS using high-resolution aerial photography and a digital photogrammetry system. *Geomorphology* 2009, *111*, 48–60.

29. Verhoeven, G.J.J.; Loenders, J.; Vermeulen, F.; Docter, R. Helikite aerial photography—A versatile means of unmanned, radio controlled, low-altitude aerial archaeology. *Archaeol. Prospect.* 2009, *16*, 125–138.

30. Eisenbeiss, H; Sauerbier, M. Investigation of UAV systems and flight modes for photogrammetric applications. *Photogramm. Rec.* 2011, *26*, 400–421.

31. Salvini, R.; Riccucci, S.; Gulli, D.; Giovannini, R.; Vanneschi, C.; Francioni, M. Geological Application of UAV photogrammetry and

terrestrial laser scanning in Marble Quarrying (Apuan Alps, Italy). In *Engineering Geology for Society and Territory—Urban Geology, Sustainable Planning and Landscape Exploitation*; Springer International Publishing: Cham, Switzerland, 2015; Volume 5, pp. 979–983.

32. Shan, B.; Luo, X.; Liang, L. Application of UAV in open-pit mine disaster monitoring. *Opencast Min. Technol.* 2013, *6*, 69–71.

33. González-Aguilera, D.; Fernández-Hernández, J.; Mancera-Taboada, J.; Rodríguez-Gonzálvez, P.; Hernández-López, D.; Felipe-García, B.; Arias-Perez, B. 3D modelling and accuracy assessment of granite quarry using unmannend aerial vehicle. In Proceedings of the ISPRS Annals of Photogrammetry, Remote Sensing and Spatial Information Sciences, Melbourne, VIC, Australia, 25 August–1 September 2012; Volume I-3, pp. 37–42.

34. McLeod, T.; Samson, C.; Labrie, M.; Shehata, K.; Mah, J.; Lai, P.; Wang, L.; Elder, J. Using video data acquired from an unmanned aerial vehicle to measure fracture orientation in an open-pit mine. *Geomatica* 2013, *67*, 163–171.

35. Harwin, S.; Lucieer, A. Assessing the accuracy of georeferenced point clouds produced via multi-view stereopsis from unmanned aerial vehicle (UAV) imagery. *Remote Sens.* 2012, *4*, 1573–1599.

36. Rosnell, T.; Honkavaara, E. Point cloud generation from aerial image data acquired by a quadrocopter type micro unmanned aerial vehicle and a digital still camera. *Sensors* 2012, *12*, 453–480. [PubMed]

37. Wang, J.P.; Gao, J.X.; Liu, C.; Wang, J. High precision slope deformation monitoring model based on the GPS/Pseudolites technology in open-pit mine. *Min. Sci. Technol. (China)* 2010, *20*, 126–132.

38. Lowe, D.G. Object recognition from local scale-invariant features. In Proceedings of the Seventh IEEE International Conference on Computer Vision, Kerkyra, Greece, 20–27 September 1999; Volume 2, pp. 1150–1157.

39. Fischler, M.A.; Bolles, R.C. Random sample consensus: A paradigm for model fitting with applications to image analysis and automated cartography. *Commun. ACM.* 1981, *24*, 381–395.

40. Kraus, K. *Photogrammetry: Geometry from Images and Laser Scans*; Walter de Gruyter: Goettingen, Germany, 2007.

41. Fraser, C.S. Digital camera self-calibration. *ISPRS J. Photogramm. Remote Sens.* 1997, *52*, 149–159.

42. Skaloud, J. Optimizing Georeferencing of Airborne Survey Systems by INS/DGPS. Ph.D. Thesis, The University of Calgary, Calgary, Alberta, 1999.

43. Kocaman, S. GPS and INS Integration with kalman Filtering for Direct Georeferencing of Airborne Imagery. Ph.D. Thesis, Institute of Geodesy and Photogrammetry, ETH Hönggerberg, Zurich, Switzerland, 2003.

44. Tong, X.H.; Hong, Z.H.; Liu, S.J.; Zhang, X.; Xie, H.; Li, Z.Y.; Yang, S.L.; Wang, W.; Bao, F. Building-damage detection using pre-and post-seismic high-resolution satellite stereo imagery: A case study of the May 2008 Wenchuan earthquake. *ISPRS J. Photogramm. Remote Sens.* 2012, *68*, 13–27.

45. Xiang, H.; Tian, L. Method for automatic georeferencing aerial remote sensing (RS) images from an unmanned aerial vehicle (UAV) platform. *Biosyst. Eng.* 2011, *108*, 104–113.

46. Lichti, D.; Gordon, S. Error propagation in directly georeferenced terrestrial laser scanner point clouds for cultural heritage recording. In Proceedings of the FIG Working Week, Athens, Greece, 22–27 May 2004.

47. Reshetyuk, Y. Self-Calibration and Direct Georeferencing in Terrestrial Laser Scanners. Ph.D. Thesis, Royal Institute of Technology, Stockholm, Sweden, 2009.

48. Harvey, B.R. Registration and transformation of multiple site terrestrial laser scanning. *Geomat. Res. Aust.* 2004, *80*, 33–50.

49. Wang, Y.; Wang, G. Integrated registration of range images from terrestrial LiDAR. *Int. Arch. Photogramm. Remote Sens. Spat. Inf. Sci.* 2008, *XXXVII* (Part B3b), 361–365.

50. Hodge, R.A. Using simulated terrestrial laser scanning to analyse errors in high-resolution scan data of irregular surfaces. *ISPRS J. Photogramm. Remote Sens.* 2010, *65*, 227–240.

51. Zhang, Y.; Gong, L.; Yan, L. Research on error propagation of point cloud registration. In Proceedings of the 2012 IEEE International Conference on Computer Science and Automation Engineering (CSAE), Zhangjiajie, China, 25–27 May 2012; Volume 2, pp. 18–21.

52. Lingua, A.; Marenchino, D.; Nex, F. Performance analysis of the SIFT operator for automatic feature extraction and matching in photogrammetric applications. *Sensors* 2009, *9*, 3745–3766. [PubMed]

53. Hartley, R.; Zisserman, A. *Multiple View Geometry in Computer Vision*, 2nd ed.; Cambridge University Press: New York, NY, USA, 2004.

54. Pesci, A.; Fabris, M.; Conforti, D.; Loddo, F.; Baldi, P.; Anzidei, M. Integration of ground-based laser scanner and aerial digital photogrammetry for topographic modelling of Vesuvio volcano. *J. Volcanol. Geotherm. Res.* 2007, *162*, 123–138.

55. Baltsavias, E.P. A comparison between photogrammetry and laser scanning. *ISPRS J. Photogramm. Remote Sens.* 1999, *54*, 83–94.

56. Wang, J.; Di, K.; Li, R. Evaluation and improvement of geopositioning accuracy of IKONOS stereo imagery. *J. Surv. Eng.* 2005, *131*, 35–42.

57. Tong, X.H.; Liang, D.; Xu, G.S.; Zhang, S.L. Positional accuracy improvement: A comparative study in Shanghai, China. *Int. J. Geogr. Inf. Sci.* 2011, *25*, 1147–1171.

58. Tong, X.H.; Liu, S.J.; Weng, Q.H. Bias-corrected rational polynomial coefficients for high accuracy geo-positioning of QuickBird stereo imagery. *ISPRS J. Photogramm. Remote Sens.* 2010,*65*, 218–226.

59. Liu, S.J.; Fraser, C.S.; Zhang, C.S.; Ravanbakhsh, M.; Tong, X.H. Georeferencing performance of THEOS satellite imagery. *Photogramm. Rec.* 2011, *26*, 250–262.

60. Henry, J.B.; Malet, J.-P.; Maquaire, O.; Grussenmeyer, P. The use of small-format and low-altitude aerial photos for the realization of high-resolution DEMs in mountainous areas: Application to the Super-Sauze earthflow (Alpes-de-Haute-Provence, France). *Earth Surf. Process. Landf.* 2002, *27*, 1339–1350.

61. Guarnieri, A.; Remondino, F.; Vettore, A. Digital photogrammetry and TLS data fusion applied to cultural heritage 3D modelling. *Int. Arch. Photogramm. Remote Sens. Spat. Inf. Sci.* 2006,*XXXVI* (Part 5).

62. Burnett, C.; Blaschke, T. A multi-scale segmentation/object relationship modeling methodology for landscape analysis. *Ecol. Model.* 2003, *168*, 233–249.

63. Yu, Q.; Gong, P.; Clinton, N.; Biging, G.; Kelly, M.; Schirokauer, D. Object-based detailed vegetation classification with airborne high spatial resolution remote sensing imagery.*Photogramm. Eng. Remote Sens.* 2006, *72*, 799–811.

64. Laliberte, A.S.; Rango, A. Image processing and classification procedures for analysis of sub-decimeter imagery acquired with an unmanned aircraft over arid rangelands. *GISci. Remote Sens.*2011, *48*, 4–23.

65. Laliberte, A.S.; Winters, C.; Rango, A. UAS remote sensing missions for rangeland applications.*Geocarto Int.* 2011, *26*, 141–156.

66. Definiens. *eCognition Developer 8.0 User Guide*; Definiens AG.: Munich, Germany, 2009.

67. Stumpf, A.; Kerle, N. Object-oriented mapping of landslides using random forests. *Remote Sens. Environ.* 2011, *115*, 2564–2577.

68. Congalton, R.G.; Green, K. *Assessing the Accuracy of Remotely Sensed Data: Principles and Practices*; CRC Press: Boca Raton, FL, USA, 2009.

69. Tong, X.; Li, X.; Xu, X.; Xie, H.; Feng, T.; Sun, T.; Jin, Y.; Liu, X. A two-phase classification of urban vegetation using airborne LiDAR data and aerial photography. *IEEE J. Sel. Top. Appl. Earth Obs. Remote Sens.* 2014.

70. Dunford, R.; Michel, K.; Gagnage, M.; Piégay, H.; Trémelo, M.L. Potential and constraints of unmanned aerial vehicle technology for the characterization of Mediterranean riparian forest. *Int. J. Remote Sens.* 2009, *30*, 4915–4935.

CITATION

Xiaohua Tong, Xiangfeng Liu , Peng Chen , Shijie Liu , Kuifeng Luan , Lingyun Li , Shuang Liu , Xianglei Liu , Huan Xie , Yanmin Jin and Zhonghua Hong, Integration of UAV-Based Photogrammetry and Terrestrial Laser Scanning for the Three-Dimensional Mapping and Monitoring of Open-Pit Mine Areas, doi:10.3390/rs70606635.

CHAPTER 2

Development of Large-Scale Land Information System (LIS) by Using Geographic Information System (GIS) and Field Surveying

Asma Th. Ibraheem

Department of Civil Engineering, College of Engineering, Nahrain University, Baghdad, Iraq

ABSTRACT

Cadastral maps are an important component of land administration in most countries. In virtually all developed countries, the needs of computerized land and geographic information systems (LIS/GIS) has given urgent impetus to computerizing cadastral maps and creating digital cadastral data bases (DCDB). This process is creating many institutional, legal, technical and administrative problems. This desire to establish DCDBs is being given increased impetus due to a new range of enabling technologies such as satellite position fixing (GPS), improved spatial data collection techniques such as digital theodolites and "soft copy" photogrammetry, as well as a vast range of new information and communications technological tools, thus contributing to the advancement and keeping up with the great countries. This paper presents the problem of cadastral maps. The hitherto existing cadastre, consisting of paper maps and land registers, is now becoming insufficient. Its shortcomings force developments leading to its improvement. One of the ways is the creation of a Land Information System. A digital cadastral map is the main component of this system. The structure and information content of the map is presented, its differences from analogue maps are shown, and the process of map creation is described. A digital cadastral map can be the basis for additional thematic layers, successively converting it into a complex system for management of administrative units.

INTRODUCTION

A cadastre is normally a parcel based, and up-to-date land information system containing a record of interests in land (e.g. rights, restrictions and responsibilities). It usually includes a geometric description of land parcels linked to other records describing the nature of the interests, the ownership or control of those interests, and often the value of the parcel and its improvements.

Graphical indices of these parcels, known as cadastral maps, show the relative location of all parcels in a given region. Cadastral maps commonly range from scales of 1:500 to 1:10,000. Large scale diagrams or maps showing more precise parcel dimensions and features (e.g. buildings, irrigation units, etc.) are often prepared by cadastral surveys for each parcel based on ground surveys and aerial photography. Information in the textual or attribute files of the cadastre, such as land value, ownership, or use, can be accessed by these unique parcel codes shown on the cadastral map, thus creating a complete cadastre [1].

The principal responsibility of the assessor is to locate, inventory, and appraise all property within the jurisdiction. A complete set of maps is necessary to perform this function. Maps help determine the location of property, indicate the size and shape of each parcel, and reveal geographic relationships that affect property value. Maps and map data are important not only for assessors, but also for other governmental agencies, the public, and the land information community (such as realtors, title companies, and surveyors). In addition, the assessor must track current ownership of all parcels, so that the proper party can receive assessment notices and tax bills. Computerization of map and parcel data can enhance the capability to manage, analyze, summarize, display, and disseminate geographically referenced information [2]. Working with digital cadastral maps and tabular parcel related data in a GIS, users can selectively retrieve and manipulate layers of parcel and spatial information to produce composite maps with only the data they need. Sharing GIS files over an internal or external data network makes parcel maps and related attribute information widely available, and reduces the duplication of effort inherent in separate map systems. Such sharing is becoming increasingly sophisticated, ranging from allowing users to download data or prepared maps, allowing users to make sophisticated queries that may draw on the power of the host GIS's software and hardware [3]. Computerized mapping systems may be referred to by several names. They include [4]:

- Geographic information system (GIS).
- Land information system (LIS).
- Digital multipurpose cadastre.
- Multipurpose land information system (MPLIS).
- Land parcel database.

THE VIEWS OF A GIS

A GIS is most often associated with a map. A map, however, is only one way that can work with geographic data in a GIS, and only one type of product generated by a GIS. A GIS can provide a great deal more problem solving capabilities than using a simple mapping program or adding data to an online mapping tool (creating a "mash-up"). A GIS can be viewed in three ways [5]:

- The Database View.
- The Map View.
- The Model View.

Together, these three views are critical parts of an intelligent GIS and are used at varying levels in all GIS applications. Data needed for GIS can be obtained in various ways and are stored in a digital form, they are known as digital data in GIS today. Digital data are obtained primarily by the following means (Ibraheem, 2008):

- Incorporating remotely sensed data into GIS.
- Digitizing existing maps and plans.
- Digitizing Ariel photographs (mono or stereo).

COMPONENTS OF A DIGITAL CADASTRAL MAPPING SYSTEM

A digital cadastral mapping system should have the following components [3]:

- Reference to a geodetic control network.
- Current base map layer (ideally, photogrammetrically derived).
- A cadastral layer delineating all real property parcels.
- Vertical aerial photographs and/or images (ideally, ortho-rectified).
- A unique parcel identifier assigned to each parcel.
- A means to tie spatial data to attribute data (ownership and parcel characteristic files).
- Additional layers of interest to the assessor, such as municipal boundaries, zoning, soil types, and flood plains.

COMPUTERISATION OF CADASTRAL MAPS

The justification for computerizing cadastral maps includes the following:

- The reduction of duplication in maintaining a cadastral base for many users.
- As a result of converting maps from one scale to another.
- To bring the cadastral map onto the same coordinate and mapping system as large scale topographic maps, thereby facilitating LIS/GIS applications.

An important issue in establishing a digital cadastral data bases (DCDB) is that computerization of the cadastral maps in general cannot be justified for land registration or land market reasons. Therefore computerization of the map requires the support of other users both financially and institutionally, [6].

At the institutional level, there is an issue of who is responsible for maintaining the DCDB and distributing the updates. Obviously it is necessary for one organization to administer the DCDB although there are various models using both government and the private sector to maintain the system [6].

LARGE-SCALE LAND INFORMATION SYSTEMS

Large-scale geographic and land information systems (GIS and LIS) are developing rapidly in local and state governments and other organizations across the country. These systems handle critical information related to land parcels, transportation, utilities, and other infrastructure and facilities. They are changing the way organizations operate and make decisions, and therefore, they affect the daily activities and lives of the citizens and customers of these organizations.

The attributes of different types of geospatial data— such as land ownership, roads and bridges, buildings, lakes and rivers, counties, or congressional districts—can each constitute a layer or theme in GIS. (See Figure 1 for a schematic representation of data layers in GIS.)

CASE STUDY

The development of the LIS of the study area goes through several distinct stages. The first is data collection and conversion, the second is editing and final is database development as well as brows and query functions through a user interface. Emphasis is placed on the method of organization determined to maximize brows and query efficiency and friendliness, by dialogue buttons, menus scripts under ArcGIS (9). This research was intended to introduce a design system to manage survey datasets through the production of GIS-ready information using appropriate standard and computing application. The trial implementation does instigate sufficient results at present stage whereby the test datasets consisting of raster image and feature classes were being managed carefully through the platform of producing and delivering GIS-ready in- formation.

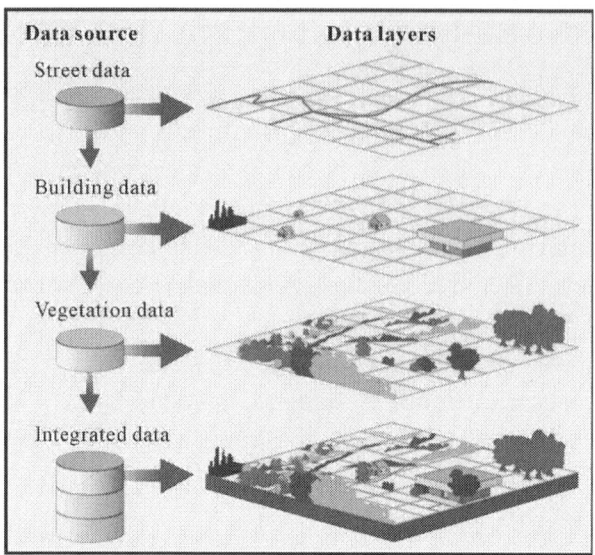

Figure 1. Example of GIS data layers or themes [7].

However there is still a need for an improved flow line of the process as more dataset type and volume covering other survey datasets held in a survey organization would be used. The testing of the design and flow line has clearly shown the possibility to disseminate, retrieve and combine those data for visualization and query over the web from multiple different data sources. ArcGIS functionality is proved offering capabilities for geospatial data interchange, manipulation and management as well. The ArcGIS application has clearly shown the successes of the concept of data integration on-the-fly from multiple heterogeneous geospatial data servers.

Data Collection and Conversion
Both analogue and digital data are gathered for the study area including:

- A digital base map for Baghdad University site.
- Eight digital Arial photos provided by surveying department as shown in **Figure 2**.
- Historical data, photographs, reproduction and surveying details.
- Additional descriptive information was also collected.

A mosaic is an assembly of series of overlapping aerial photographs to form one continuous picture of the terrain. It may consist of a single strip of photographs, termed a strip mosaic, or it may be contain many overlapping strips. The images used in this study captured from space by the military in (July 1985) with focal length (152.16 mm), flying height (456.48 m) and scale the image (1/ 3000) which is illustrated in **Figure 2**. The completion of this work needing to several enough aerial photos of the area. The mosaics have been found that the University of Baghdad covered pace aviation are flight (110, 111). All the airline has four images so that there is a common area shows each image of the photo that followed in the line of flight, one is called the forward overlap or overlap the front and the amount (60%). And also there is an overlap (side lap) or (End Lap) hereunder (30%) were converted this image to a digital format (Digital form) using (Scanner). Tthen saved on CD in the Department of Surveying Engineering, which is used to transfer this image to a calculator to work with them.

Matching the mosaic images usually implies that the radiometric intensity data from one image representing a particular feature must be matched to the intensity data from the second image, representing the same feature. This implies more than just matching image intensity data correlation, because the same piece feature may look considerably different radio metrically from different point of view, or at different time. **Figure 3** shows the matching the mosaic of eight digital Arial photos of Baghdad University site by using ERDAS LPS (V 9.2) software.

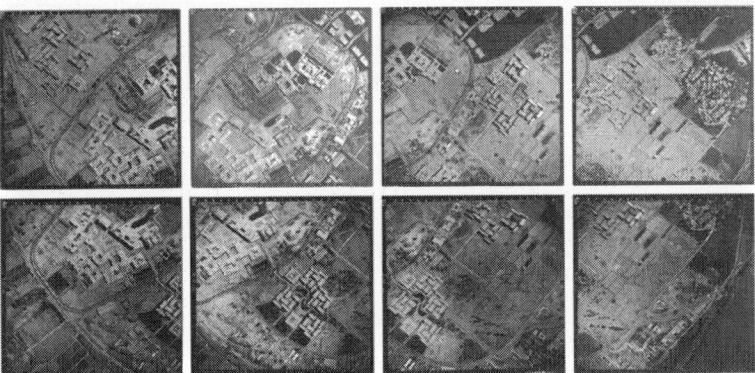

Figure 2. Eight digital Arial photos of Baghdad University site.

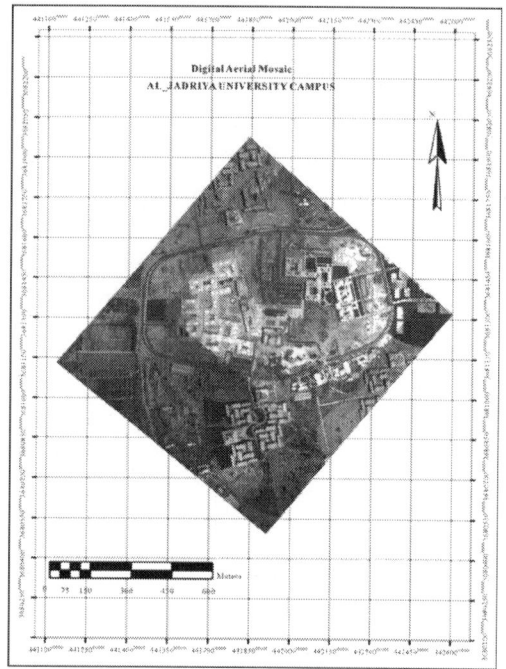

Figure 3. Matching the mosaic of eight digital Arial photos of Baghdad University site by using ERDAS LPS software.

Lis Structure and Analysis

The LIS's structure is performed into the ArcGIS environment, which provides easy and efficient data management and analysis. The specific application includes the integration of the above mentioned coverage into ArcGIS as themes and the implementation of the database, and organized according to the following thematic data types (**Figure 4**):

- Colleges;
- Departments;
- Roadways;
- Parks;
- Green spots;
- Channels;
- Mosque;
- Sport Fields;
- Residents.

Zooming in the map, a more detailed display of the area is obtained, facilitating the investigation process. Through the use of the menu, the desired entity type is displayed, further divided into sub-types, according to specific characteristics. The Start-up screen of the system displays a map of the whole area included in the ArcGIS software (**Figure 5**). An error of less than 1.0000 pixels is acceptable. An error of greater than 1.0000 indicates that the points were inaccurately measured or poorly identified. But in this study the resulted accuracy Root Mean Square Error (RMSE) was (0.0775) which was very acceptable.

The ArcGIS software was used to provide a digital map with multi layers for that study area. The layers which were formed were eight layers are: layer of colleges, layer of departments, layer of roads way, layer of mosque, layer of parking, layer of sport fields, layer of gardens, layer of open channel, layer of sub roads and layer of status. Figures (6)-(8) illustrate the drawings of some of these layers (roads way, gardens and open channel).

Figure 4. The zoom in the entity of College of Engineering in Baghdad University site by using ArcGIS environment.

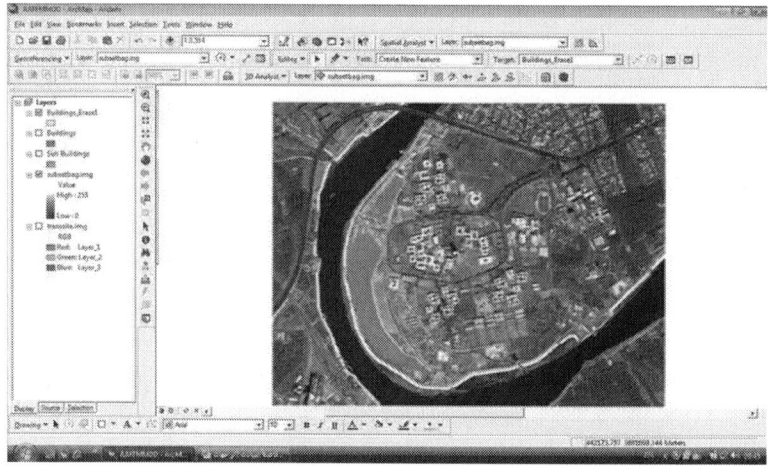

Figure 5. A thematic map of the whole area included in the ArcGIS software of College of Engineering in Baghdad University site.

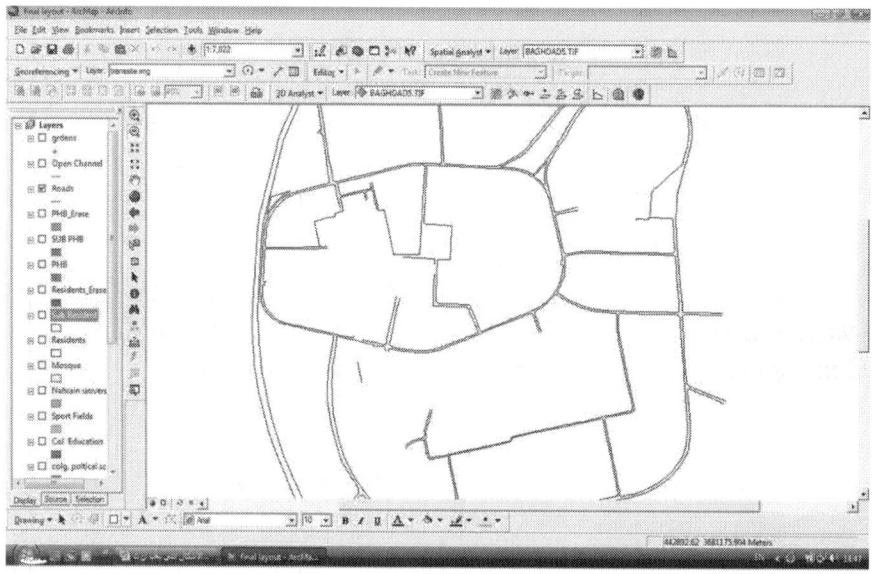

Figure 6. Layer of road way.

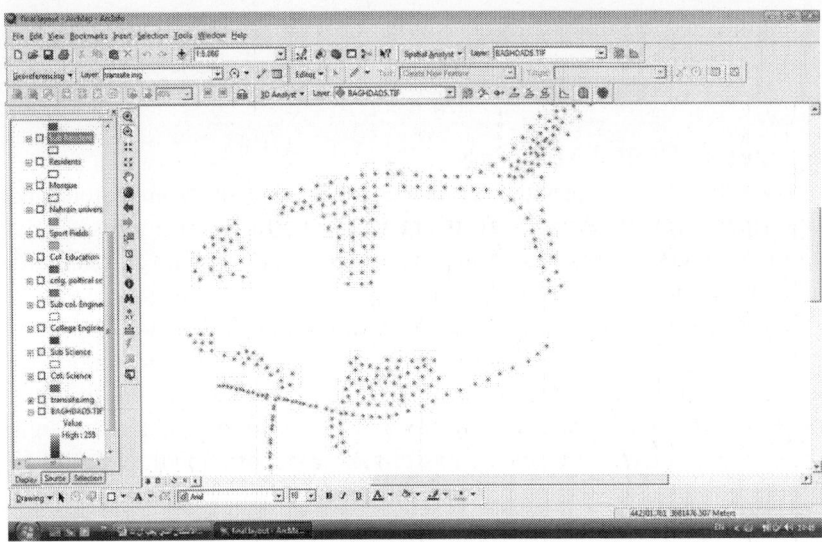

Figure 7. Layer of gardens.

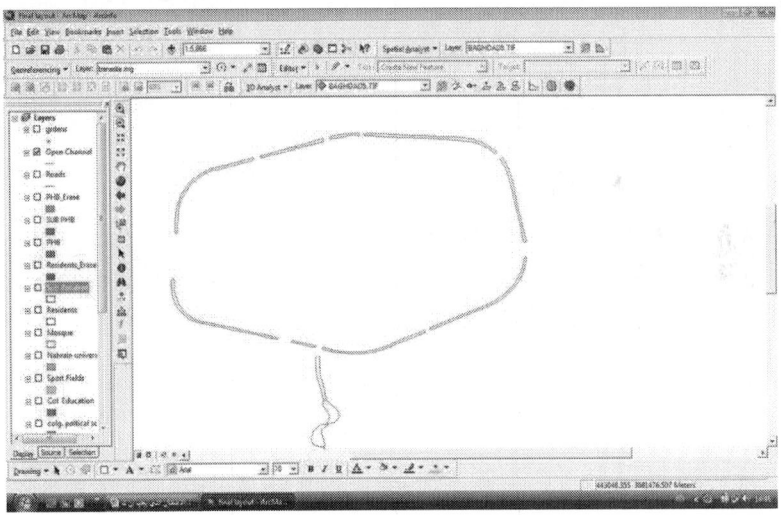

Figure 8. Layer of open channels.

Clicking on the area of the interest a second dialog window is presented to the user including all the entity types of the area (Figures 9(a)-(f)). A button linked to an Arial photo of the area is also available.

The application results in digital dynamic map of 1/ 10,000 scale, as shown in **Figure 10**, associated to descriptive and multimedia information about the site of Baghdad University.

Accuracy of Digital Mapping

In this study, after the production of mosaic for that study area stereo pair of the final form the ArcGIS software was used to prepare a digital map with multi layers. The layers include roads layer, colleges layer, residential sectors layer, sport fields layer, parking layer, sub roads layer, gardens layer and urban places layer. After the completion of the drawing layers, the digital map was produced in its final form with scale of 1:10,000. This scale is appropriated to the scale of digital aerial photographs that were used in the production, which in turn was to be used to produce the digital map; thirteen points were selected as check points as shown in **Table 1** to compute the resulted accuracy Root Mean Square Error (RMSE) by using the following equations:

Table 1. Ground control points (GCPs) in UTM system.

Remark	Y (Easting)	X (Northing)	Sta.
4U.B.	3681323.884	441959.883	
1	81531.108	41509.005	
2	81553.005	41676.445	
3	81591.506	41966.536	
8	81505.107	42321.602	UTM Coordinate of Station 4U.B. the National Grid.
9	81233.677	42354.858	
10	81117.789	42067.319	
11	81071.343	41858.940	
12	81109.611	41702.700	4U.B. (4) University Building.
P1	81184.942	42547.451	
19	80536.217	42093.680	
20	80542.393	41904.364	
21	80706.598	41734.572	

$$R_i = \sqrt{Rx_i^2 + Ry_i^2}$$

(1)

where R_i: The RMSE for check point (i).

Rx_i: The X residual for check point (i), (the distance between the source and the transformed coordinates in x

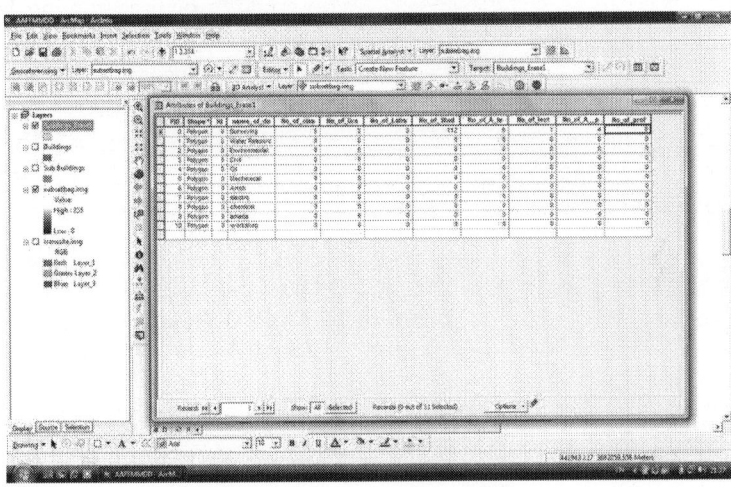

(a)

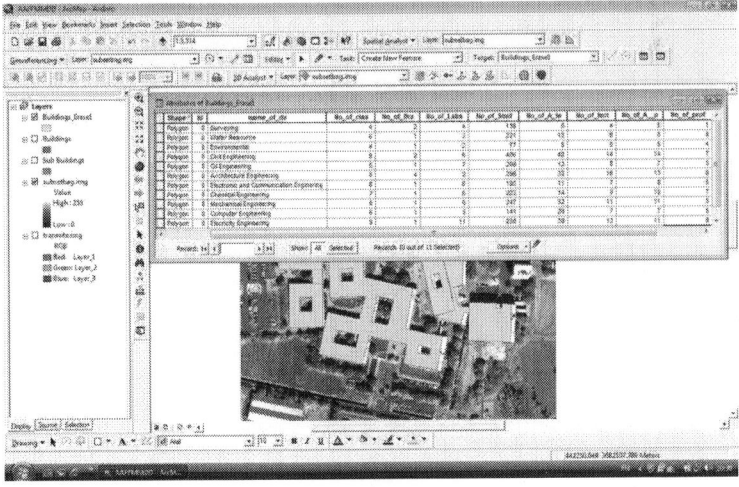

(b)

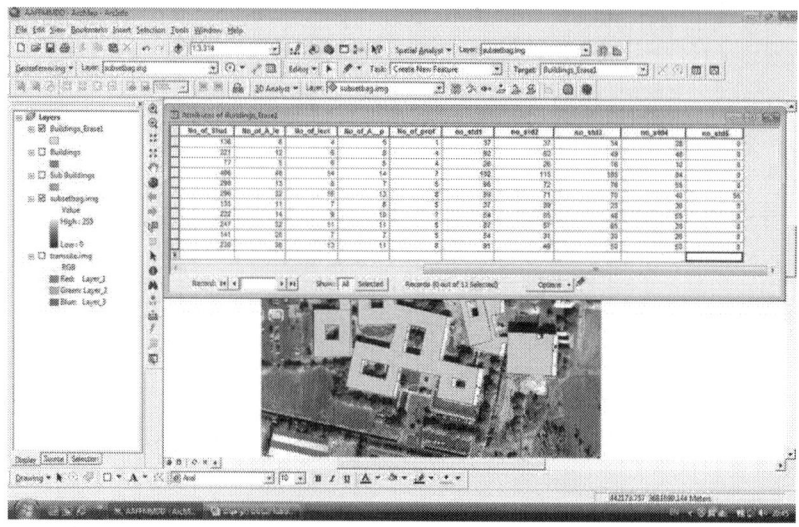

(c)

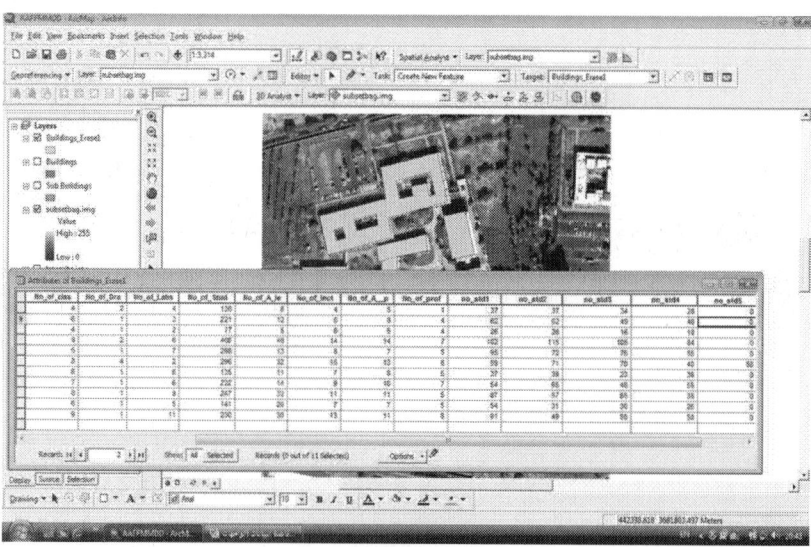

(d)

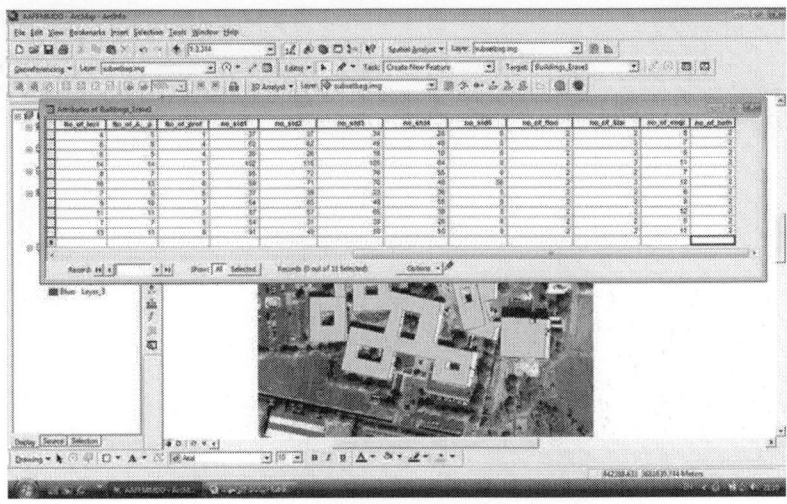

(e)

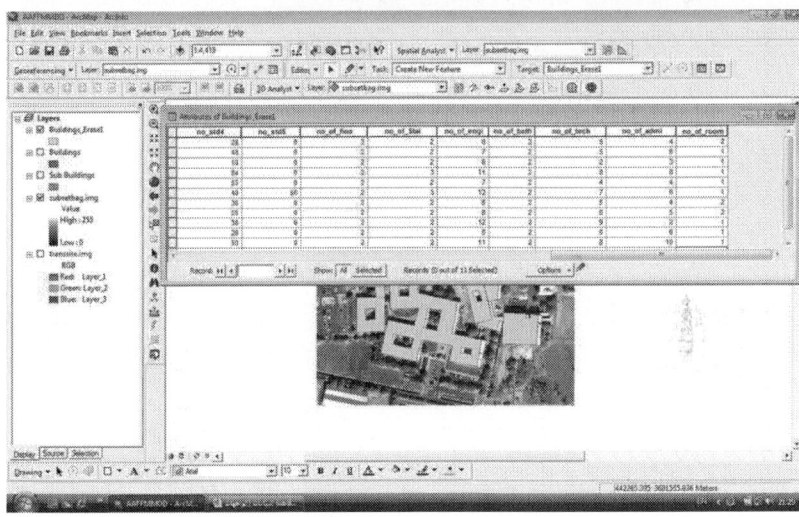

(f)

Figure 9. Dialog windows of the database included in the ArcGIS software of College of Engineering in Baghdad University site.

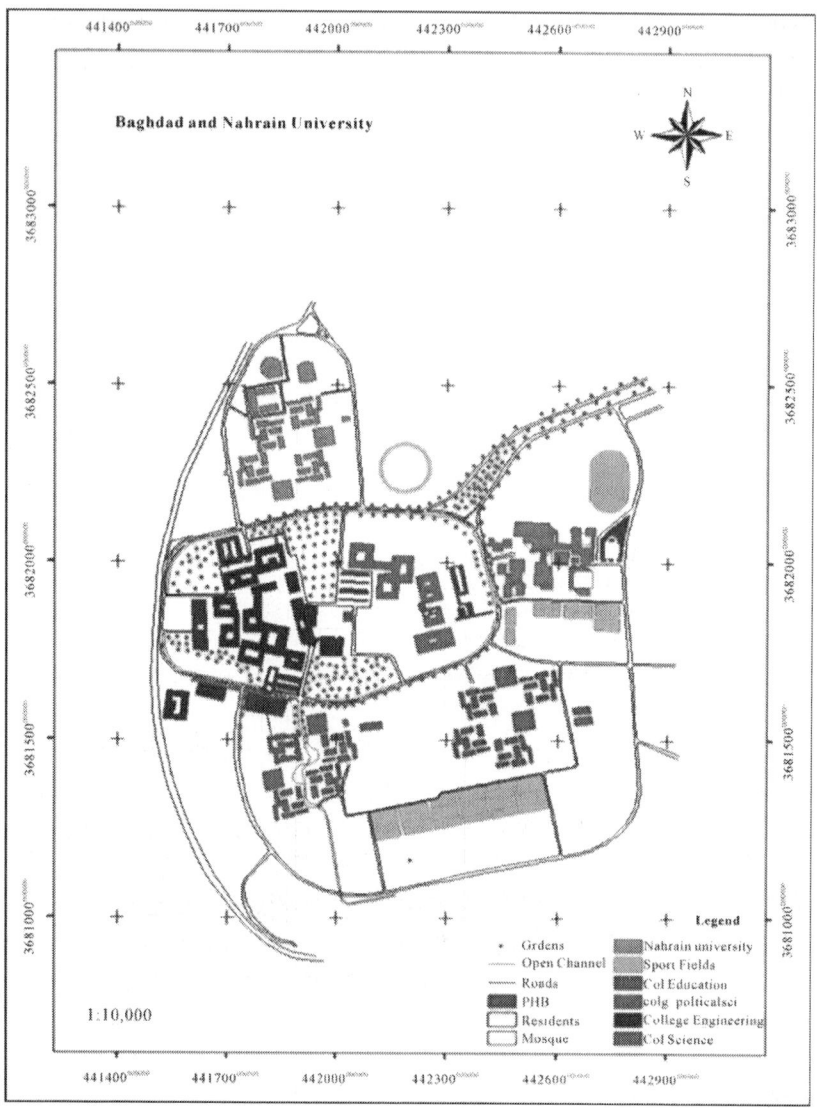

Figure 10. The digital cadastral map of Baghdad University site by using ArcGIS environment, scale 1:10,000.

direction).

Ry_i: The Y residual for check point (i), (the distance between the source and the transformed coordinates in y direction).

Depending upon the residuals, the RMSE in X coordinate, the RMSE in Y coordinate, and the total RMSE can be computed from the following equations:

$$R_x = \sqrt{\frac{1}{n}\sum_{i=1}^{n} R_{xi}^2}$$

(2)

$$R_y = \sqrt{\frac{1}{n}\sum_{i=1}^{n} R_{yi}^2}$$

(3)

$$R_T = \sqrt{R_x^2 + R_y^2}$$

(4)

where:
R_T: total Root Mean Square Error (RMSE).
n: number of check points.

Thirteen point were selected for the checking the accuracy in digital map, the resulted accuracy ((Root Mean Square Error (RMSE)) computed by using a special Equation (4) as described above was 55 cm, which was suitable for the production of large scale mapping. The selected scale map is suitable and appropriate for the scale of aerial photographs that have been used in the production of the map, and the possible use of the map produced and used at any time possible and updates its data using Geographic Information System (GIS) software.

CONCLUSIONS

It is well known that GIS has already become a standard tool for handling spatial data. GIS are now more commonplace in Iraq. Adding multimedia material and organizing friendly entries and alternative search techniques, the system then becomes an efficient tool for different users. Until only a few years ago the map was perceived as a static, plain view of preselected areas, available at fixed scales and, due to the development of the landscape, often out of date. Now, it is evolving into a dynamic, continually updated network of interrelated databases with volumes of geographically referenced information linked to a comprehensive digital cadastral database. As shown above, our study results in efficient Land Information System for the study area. This system may be improved, by adding the following steps:

- Promote and ensure the reliability and integrity of large-scale land information systems and Facilitate collaboration between GIS and surveying professionals at the local, regional, and national levels.
- Promote the use of sound surveying and mapping principles in the development and the use of land information systems.
- Foster the development and adoption of useful standards, specifications, and procedures for the development, operations, and the use of land information systems.
- The LIS's structure conducts the user's navigation through alternative searching paths, created by the application.

REFERENCES

1. J. L. G. Henssen, "Cadastre, Indispensable for Development," International Institute for Aerospace Survey and Earth Sciences (ITC), Enschede, 1990.
2. G. Larsson, "Land Registration and Cadastral Systems," Longman Scientific and Technical, New York, 1991.
3. W. Y. Wan and I. P. Williamson, "A Review of the Digital Cadastral Databases in Australia and New Zealand," The Australian Surveyor, Vol. 40, No. 1, 1995, pp. 41-52.
4. ASCE Editorial, "Basic of Geographic Information System," Journal of Computing in Civil Engineering, ASCE, January 1998, pp. 1-4.
5. A. Th. Ibraheem, "The Application of Geographical Information System in Civil Engineering," Integrating Teaching and Research with Community Service, Book No. 87, University of Sharjah, Sharjah, 2008, pp. 436- 455.
6. I. Williamson and S. Enemark, "Cadastre and Land Management," The University of Melbourne, Melbourne, 1994.
7. P. Folger, "Geospatial Information and Geographic Information Systems (GIS): Current Issues and Future Challenges," CRS Report, Washington DC, 2009.

CITATION

A. Ibraheem, "Development of Large-Scale Land Information System (LIS) by Using Geographic Information System (GIS) and Field Surveying," Engineering, Vol. 4 No. 2, 2012, pp. 107-118. doi: 10.4236/eng.2012.42014.

CHAPTER 3

Advancing of Land Surface Temperature Retrieval Using Extreme Learning Machine and Spatio-Temporal Adaptive Data Fusion Algorithm

Yang Bai [1,2], Man Sing Wong [1], Wen-Zhong Shi [1], Li-Xin Wu [2] and Kai Qin[2]

[1]Department of Land Surveying and Geo-Informatics, The Hong Kong Polytechnic University, Kowloon, Hong Kong;
[2]School of Environment Science and Spatial Informatics, China University of Mining and Technology, Xuzhou 221116, China;

ABSTRACT

As a critical variable to characterize the biophysical processes in ecological environment, and as a key indicator in the surface energy balance, evapotranspiration and urban heat islands, Land Surface Temperature (LST) retrieved from Thermal Infra-Red (TIR) images at both high temporal and spatial resolution is in urgent need. However, due to the limitations of the existing satellite sensors, there is no earth observation which can obtain TIR at detailed spatial- and temporal-resolution simultaneously. Thus, several attempts of image fusion by blending the TIR data from high temporal resolution sensor with data from high spatial resolution sensor have been studied. This paper presents a novel data fusion method by integrating image fusion and spatio-temporal fusion techniques, for deriving LST datasets at 30 m spatial resolution from daily MODIS image and Landsat ETM+ images. The Landsat ETM+ TIR data were firstly enhanced based on extreme learning machine (ELM) algorithm using neural network regression model, from 60 m to 30 m resolution. Then, the MODIS LST and enhanced Landsat ETM+ TIR data were fused by Spatio-temporal Adaptive Data Fusion Algorithm for Temperature mapping (SADFAT) in order to derive high resolution synthetic data. The synthetic images were evaluated for both testing and simulated satellite images. The average difference

(AD) and absolute average difference (AAD) are smaller than 1.7 K, where the correlation coefficient (CC) and root-mean-square error (RMSE) are 0.755 and 1.824, respectively, showing that the proposed method enhances the spatial resolution of the predicted LST images and preserves the spectral information at the same time.

INTRODUCTION

In order to monitor the rapid and continual changes of the global environment, Land Surface Temperature (LST), as the prime and basic physical parameter of the earth's surface, has been studied for over a decade. LST plays a key role in modeling the surface energy balance [1,2] and has a significant impact on analyzing the heat-related issues such as soil moisture [3,4], evapotranspiration [5,6,7], and urban heat islands [8,9]. Compared with traditional methods using data from weather stations, remote sensing satellite images provide a more effective and efficient method to estimate LST and offer a synoptic view of the study area. However, due to the limitations on both spatial and temporal resolution of the existing satellite sensors, there is no earth observation which can obtain Thermal Infra-Red images (TIR) at detailed spatial- and temporal-resolution simultaneously.

Currently, the applications of thermal infrared remote sensing in urban environment studies require heat-related information at high spatial resolution, as well as high temporal resolution [7]. However, in order to collect more reflected and emitted signal from the earth, large spatial coverage, e.g., lower spatial resolution from the earth observation is required. For instance, several sensors such as Landsat Thematic Mapper (TM)/Enhanced TM Plus (ETM+)/Operational Land Imager (OLI) and Advanced Space borne Thermal Emission and Reflection Radiometer (ASTER), obtain TIR data between 60 m and 120 m [10,11,12,13] and they are always used for regional and global LST studies [14,15]. In addition, the platforms with a minimum 16-day revisit cycle may prohibit its application. The revisit cycle for a particular area may be extended due to the poor atmospheric conditions, such as cloud and haze [16]. Particularly in cloud-prone environments, e.g., Hong Kong and the Pearl River Delta region, the poor atmospheric conditions result in a very low probability of obtain cloud-free Landsat imagery (e.g., 10% in a year with cloud cover below 10%) [17].

On the contrary, other sensors, such as Advanced Very High Resolution Radiometer (AVHRR), and MODerate resolution Imaging Spectroradiometer (MODIS), provide a daily revisit cycle, but at a coarser spatial resolution ranging from 250–1000 m, which may not be able to detect the high level of detailed information and seriously impede their potential applications [18,19]. As a consequence, it is necessary to develop a new image fusion method that can integrate complementary characteristics from multi-sensors, in order to generate synthetic LST data with high spatial and temporal resolution.

There are some previous studies of remote sensing data fusion methods using multiple optical sensors [20,21,22]. Considering different characteristics between the TIR band and the visible bands, traditional image fusion models, such as the widely used Principle Component Analysis (PCA) based fusion methods [23,24], Intensity-Hue-Saturation (HIS) transformation method [25,26] and wavelet-based image fusion methods [27,28], are more concentrating on the visual effects of the fused images, which may not be useful for quantitative remote sensing applications. Hence, some downscaling approaches have been developed for TIR images. A generalized theoretical framework with semi-empirical regression and modulation integration techniques was constructed by Zhan et al. [29] and they evaluated all three fusion data including digital number, radiance and land surface temperature. Rodriguez Galiano et al. [30] applied the Downscaling CoKriging (DCK) method to obtain high spatial resolution LST images using the same scene data at a coarser resolution and the Normalized Difference Vegetation Index (NDVI). The downscaled TIR band was also used to improve the land cover classification accuracy and derive evapotranspiration images of surface energy balance model in a large heterogeneous landscape [31,32,33].

However, this study is not only aiming on downscaling the TIR images to retrieve higher details of spatial information, but also on fusing high spatial and high temporal data from multi-sensors, e.g., blending high spatial resolution Landsat data and high temporal MODIS data, in order to generate synthetic LST maps. A spatial and temporal data fusion model named the Spatial and Temporal Adaptive Reflectance Fusion

Model (STARFM) [34] was firstly developed to predict daily surface reflectance at a Landsat 30 m spatial resolution using one or more pairs of Landsat and MODIS images on the same day, and one MODIS image on the predicted day, based on the weighted average approach. The STARFM algorithm has been widely used to provide information for monitoring the seasonal changes in vegetation cover and large changes in land use [34]. The performance of STARFM is highly dependent on the characteristic of the landscape patch size. In order to overcome the shortcomings of STARFM, Zhu et al.[35] proposed an Enhance STARFM model to improve the accuracy of the predicted image. Compared to STARFM, ESTARFM results preserve more spatial details on finer resolution images, especially for heterogeneous land covers. Hilker et al. [36] also designed a new fusion approach named Spatial and Temporal Adaptive Algorithm for mapping Reflectance Change (STAARCH) for distinguishing and recording the changes of reflectances in both MODIS and Landsat data. Among the above approaches, only STAARH considers the change in land cover types, but it can only be used for predicting the changes in forest [36]. A semi-physical fusion approach employs the MODIS BRDF/Albedo land surface characterization product and Landsat ETM+ data to predict ETM+ reflectance, and the method can also be used for ETM+ cloud/cloud shadow, SLC-off gap filling, and relative radiometric normalization [37]. Through an unmixing-based spatial and temporal fusion model, Zurita Milla et al. [38] integrated Landsat TM and MERIS data, however, the results were highly dependent on the quality of the land cover map generated from Landsat images. In addition, a robust SParse representation-based SpatioTemporal reflectance Fusion Model (SPSTFM) has been developed to predict high spatial resolution surface reflectance through data blending with low spatial resolution data. Results show that the SPSTFM is able to capture the changes of surface reflectances in both the changes of phenology and land-cover types [39,40]. These methods were designed for data fusion of surface reflectance rather than for TIR data. Huang et al. [41] developed a spatiotemporal image fusion model based on bilateral filtering to derive high resolution LST images. This model works in a densely time-series LST data for urban heat island monitoring. Weng et al. [42] proposed a new data fusion model named Spatio-Temporal Adaptive Data Fusion

Algorithm for Temperature Mapping (SADFAT) to predict synthetic thermal radiance and LST images at both high temporal and spatial resolution by blending daily MODIS and periodic Landsat datasets. The prediction, accuracy of SADFAT, as measured by the mean absolute difference, ranged from 1.25 K to 2 K. SADFAT was developed for predicting Landsat-like thermal radiance and LST data based on the Spatial and Temporal Adaptive Reflectance Fusion Model (STARFM) which also considers annual temperature cycle (ATC) and urban thermal landscape heterogeneity [42].

In this paper, in order to generate synthetic high spatial resolution LST images from available satellite-borne sensors, a thermal sharpening method based on extreme learning machine (ELM) algorithm for neural network regression model was firstly adopted, in order to enhance the 60 m Landsat ETM+ TIR band to 30 m resolution. Then, the MODIS LST and enhanced Landsat ETM+ TIR data were blended through Spatio-temporal Adaptive Data Fusion Algorithm for Temperature mapping (SADFAT) (followed by the algorithm developed by Weng *et al.* [42]) to generate synthetic LST data at 30 m resolution.

METHODOLOGY

Extreme Learning Machine (ELM) Algorithm

For pattern recognition and image classification, traditional image fusion methods focus on the enhancement of visual effects, including color and texture enhancement; however, for quantitative remote sensing applications and thermal infrared image analysis, the preservation of spectral brightness values is a higher priority. In this paper, an extreme learning machine (ELM) algorithm [43,44] was selected to obtain the internal physical regression relationship between the 60 m Landsat ETM+ TIR image and corresponding 30 m multispectral bands (including visible, near and shortwave infrared bands).

Extreme Learning Machine (ELM) proposed by Huang *et al.* [45] is designed for single-hidden layer feed-forward neural networks (SLFNs) that can adjust the input weights and determine the output weights analytically. Due to the random determination of the input weights and hidden biases, ELM requires numerous of hidden neurons. In practice, the number of hidden neurons should be larger than the number of the variables in dataset, since the useless neurons from the hidden layer will be pruned automatically.

Figure 1 shows the structure of single hidden layer feed forward neural network based on ELM using the activation function, $g(x)=sig(w_i \cdot x_i + bi)$.

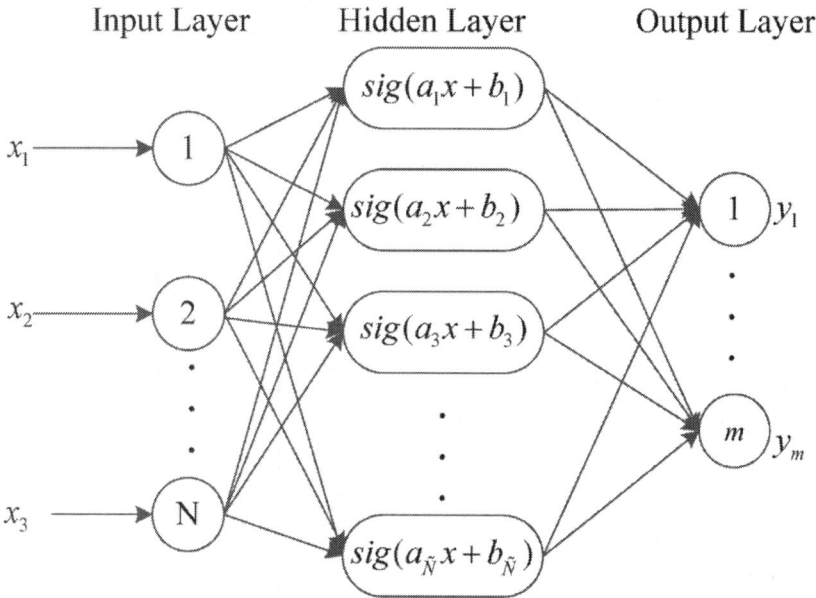

Figure 1. Structure of the single hidden layer feed forward neural network using Extreme Learning Machine (ELM).

The output weight β for a training set $A=\{(x_i, t_i)|x_i \in Rn, t_i \in Rm, i=1,2,\cdots,N|\}$ with activation function $g(x)$ and hidden neuron $N^\sim$ can be calculated according to the following steps:

(i). Allocate the input weight wi and bias bi,i=1,2,$\cdots$,N$^\sim$ arbitrarily.
(ii). Compute the hidden layer output matrix H.
(iii). Use the equation $\beta=H-1T$ to calculate the output weight matrix β.

Spatio-Temporal Adaptive Data Fusion Algorithm for Temperature Mapping (SADFAT)

The Spatio-Temporal Adaptive Data Fusion Algorithm for Temperature Mapping (SADFAT), developed by Weng *et al.* [42], was used to predict synthetic Landsat-like thermal radiance and LST images in this paper. In the algorithm, the MODIS radiance images were first resampled to the same spatial resolution (e.g., 30 m) of Landsat ETM+ images, and M is denoted as the MODIS pixel while L is denoted as the Landsat pixel. In urban areas, most of the pixels from MODIS images covered more than one land cover type, and these were named mixed pixels. Therefore, the linear spectral mixture analysis (LSMA) algorithm was employed to define the radiance of a mixed pixel. In order to correlate the Landsat radiance with the MODIS radiance, it is assumed that each L pixel is deemed as pure pixel and can be considered as an end-member of a M pixel; as a result, the radiance of the M pixel can be defined as:

$$R_M(t_j) = \sum_{i=1}^{N} l_i R_M(x, y, t_j) = \sum_{i=1}^{N} l_i \left(\frac{1}{k_1} R_L(x, y, t_j) - \frac{k_2}{k_1} \right) \quad (1)$$

where R denotes the radiance at the satellite sensor, N is the pixel number of Landsat within a MODIS pixel and l_i denotes the fraction of each Landsat pixel, (x, y) represents a given location, t is the acquisition date, and k_1, k_2 are the coefficients for the relative adjustment for the Landsat and MODIS radiance pixels [42]. Therefore, if there are two pairs of Landsat ETM+ and MODIS image acquired at t_1 and t_2, respectively, Equation (2) indicates that the ratio of the radiance change of j_{th} L pixel to the radiance of corresponding M pixel is constant for a certain L pixel can be quantified as below:

$$\frac{R_{jL}(t_2) - R_{jL}(t_1)}{R_M(t_2) - R_M(t_1)} = \frac{\cos(\omega_k + a\bar{e})}{\frac{1}{k_1} \sum_{i=1}^{N} l_i \cos(\omega_i + a\bar{e})} = h_j$$

$$(2)$$

where h_j is denoted as the conversion coefficient for the purpose of consistency, ω reflects the phase shift of a pixel and is related with the thermal characteristics of land cover, and it is a constant if the land surface materials does not change in the period of observation [42]. Thus, if there is a pair of L and M radiance images at t_0 and an M radiance image at t_p, the L radiance image at t_p can be predicted by the following equation:

$$R_L(x,y,t_p) = R_L(x,y,t_0) + h(x,y) \times [R_M(x,y,t_p) - R_M(x,y,t_0)]$$

(3)

By adding information from neighbouring spectral similar pixels, a moving window would be used to calculate the radiance of the central pixel [34]. Therefore, supposing s is the moving window size, the predicted L pixel radiance can be rewritten as:

$$R_L(x_{s/2}, y_{s/2}, t_p) = R_L(x_{s/2}, y_{s/2}, t_0) + \sum_{i=1}^{N} W_i \times h_i \times [R_M(x_i, y_i, t_p) - R_M(x_i, y_i, t_0)]$$

(4)

where W_i represents the weight of a neighboring similar pixel, and N is the number of the spectral similar pixel. In this paper, after predicting the radiance image using SADFAT method, enhanced Landsat ETM+ TIR data with both high spatial and temporal resolution would be converted to LST using the generalized single channel method [46]. More details of SADFAT can be referred to Weng *et al.* [42].

Implementation of the Proposed Data Fusion Model

Figure 2 presents a flowchart of the proposed fusion model. The implementation consists of two steps to produce synthetic spatial resolution LST maps using Landsat ETM+ and MODIS LST images. In the first step, a thermal sharpening method using ELM algorithm was selected to improve the Landsat ETM+ TIR 60 m resolution image to 30 m resolution. The multispectral and TIR digital number (DN) values were first converted to the radiance values using metadata, and these could ensure the two satellite images have a strong intrinsic correlation of the same land surface types. Then, the hidden layer output matrix,*i.e.*, internal physical regression relationship between the 60 m spatial resolution of Landsat ETM+ TIR band and corresponding six 30 m resolution multispectral bands, was ascertained through the ELM algorithm where the hidden neurons were set to 1000. Finally, the enhanced 30 m resolution Landsat TIR data were computed using the weight vector obtained by neural network regression model.

In the second step, the MODIS LST and enhanced Landsat ETM+ TIR data were blended using SADFAT algorithm [42] to generate a final synthetic LST data at 30 m resolution. The algorithm requires at least

two pairs of Landsat and MODIS images acquired at the same date and a set of MODIS images on the desired prediction dates. Before implementing the SADFAT algorithm [42], MODIS LST images should be converted to radiances at the Landsat ETM+ effective thermal wavelength and all the images including enhanced Landsat ETM+ TIR and the converted MODIS radiance images should be geo-registered to the same coordinate system and atmospherically calibrated to the surface radiance. The inputs of this step are two pairs of L and M images at t_1 and t_2 and one M image at the prediction date t_p.
The details of data processing are described as below:

i.   Two L images were used to search for the spectrally similar pixels using the method described by Gao et al. [34] that define a difference threshold between the central pixel and the neighbouring pixels in a moving window.

ii. The combined weight and conversion coefficient for each similar pixel were computed. Here, a similar pixel with higher thermal similarity and shorter distance to the central pixel would yield a higher weight and the conversion coefficients were decided by the regression analysis of the similar pixels [35].

iii. Equation (5) was employed to compute the desired predicted image at t_p. Considering the temporal weights of the two images given by the temporal changes in coarser radiance images, an accurate radiance image can be computed by using the weighted combination of the two predicted radiance images as follows:

$$L(x_{s/2}, y_{s/2}, t_p) = T_{t_1} \times P_{t_1}(x_{s/2}, y_{s/2}, t_p) + T_{t_2} \times P_{t_2}(x_{s/2}, y_{s/2}, t_p)$$

(5)

where $(x_s/2, y_s/2)$ represents a location of the central pixel in a moving window, t is the acquisition date, P_{t1} and P_{t2} are the predicted radiance image using L image at t_1 and t_2 as the base image, respectively, in which the temporal weight T_k can be calculated as:

$$T_k = \frac{1/(\sum_{i=1}^{s}\sum_{j=1}^{s} M(x_i, y_j, t_k, B) - \sum_{i=1}^{w}\sum_{j=1}^{w} M(x_i, y_j, t_p, B))}{\sum_{k=t_1,t_2} 1/(\sum_{i=1}^{s}\sum_{j=1}^{s} M(x_i, y_i, t_k, B) - \sum_{i=1}^{w}\sum_{j=1}^{w} M(x_i, y_j, t_p, B))}$$

(6)

where $M(x, y, t, B)$ is the resampled radiance at time t of band B.

iv.   Finally, the LST images can be derived using the generalized single channel method.

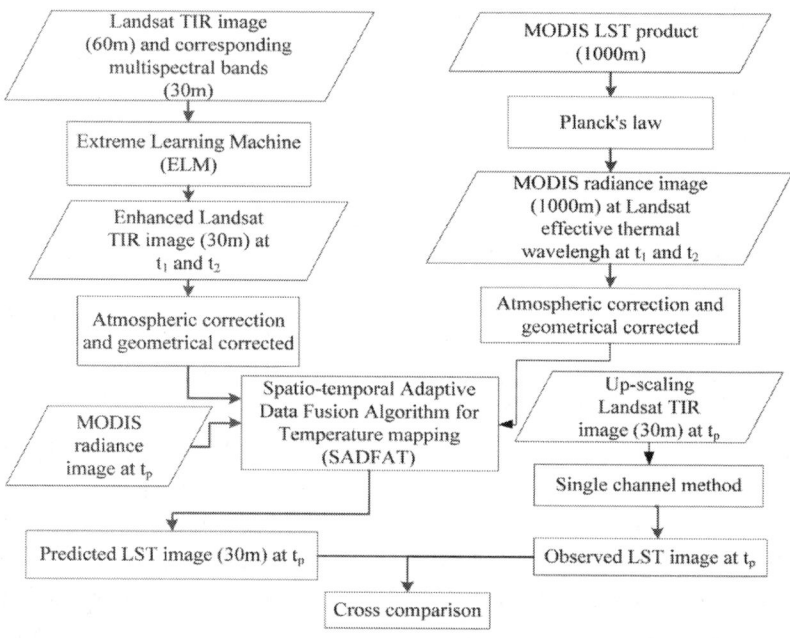

Figure 2. Flowchart of the proposed fusion model for predicting synthetic LST image at 30 m resolution.

RESULTS

Study Area

A study area of 12 km × 12 km, part of Guangzhou, China, was selected (see Figure 3). The study area consists of various landscapes, including water, impervious surface, bare soil, vegetation, *etc.*, and possesses a subtropical climate with moist summers and dry winters.

Three Landsat ETM+ images covering the study area (path/row: 122/44) were obtained from the United States Geological Survey (USGS) website. These datasets were acquired on 20 October, 7 December and 23 December 2013, and they are the L1G level product and are geographically corrected.

The corresponding daily MODIS LST (MOD11A1), reflectance data (MOD09GA) and water vapour content data (MOD05_L2) were downloaded through the Atmosphere Archive and Distribution System website. The MODIS LST (MOD11A1) products were derived from the generalized split-window LST algorithm and the uncertainty of the MODIS-LST production algorithm is around 1 °C in the range of −10 °C to 50 °C, for the surfaces with known emissivities [47].

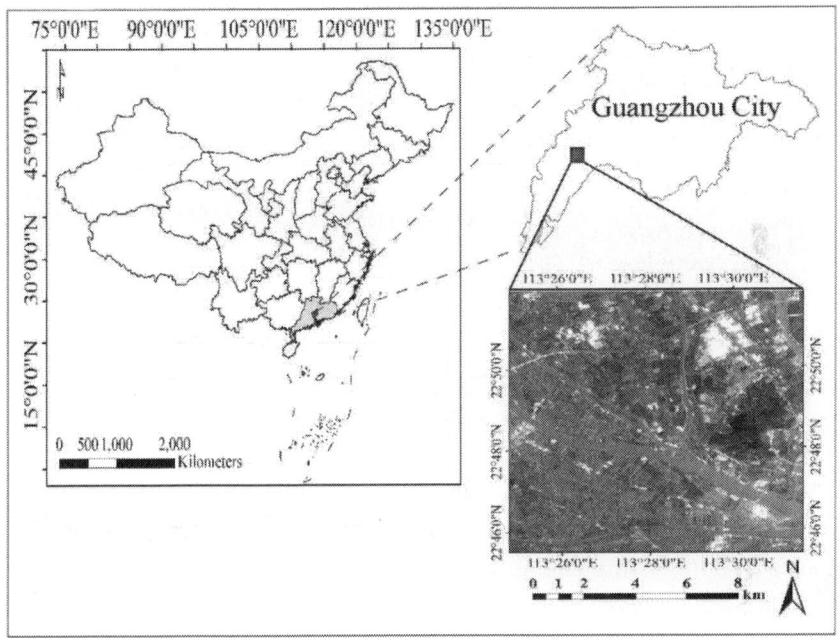

Figure 3. Location of the study area.

Experiment Results

In this study, the developed fusion methodology was applied to three pairs of Landsat ETM+ and MODIS images both in testing and simulated experiments (Figure 4). Two pairs of the MODIS and Landsat images, acquired on 20 October and 23 December 2013, respectively, were used as base images input to the fusion model, while the remaining images acquired on 7 December 2013, was used to validate the accuracy of prediction. For the testing experiment, the original multispectral images and TIR image of Landsat ETM+ were downscaled to 120 m using the pixel averaging method, then the ELM algorithm was employed to enhance the spatial resolution of the degraded TIR image from 120 m to 60 m, and the original TIR image at 60 m was used as the referenced data for validation. The correlation coefficient (CC), root-mean-square error (RMSE), average difference (AD) and absolute average difference (AAD) were selected as indicators to evaluate the accuracy of thermal spatial sharpening of Landsat TIR images using ELM algorithm and to validate the synthetic LST. For the simulated experiments, since there is no reference to TIR images at 30 m resolution, the reliability of ELM and SADFAT was validated only through visual effects.

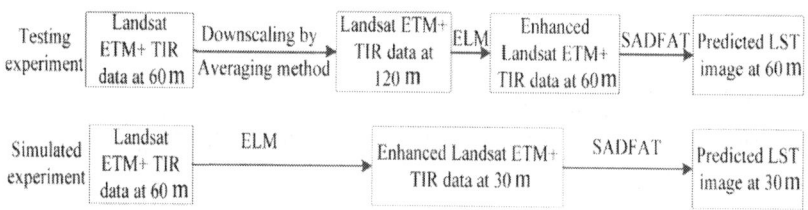

Figure 4. The flowchart of the testing and simulated experiment.

Figure 5 and Figure 6 showed the thermal spatial sharpening results (lower row) of all the testing and simulated Landsat radiance images (middle row) using the extreme learning machine (ELM) algorithm for the neural network regression model. Comparing the spatial sharpened images in both testing and simulated experiments, it can be observed that the sharpened images have more spatial details, such as roads, textural information of forest, urban streets and edges of rivers. Additionally, CC, RMSE, AD, AAD were computed between the sharpened and original radiance images at both 60 m resolution (Table 1). The RMSEs between the sharpened and original radiance images are 0.0844, 0.0891, and 0.0909, and the CC values are 0.8788, 0.8251, and 0.8017 on different dates, respectively. AD values on 20 October and 7 December 2013 are negative values showing that the predictions overestimate the TIR data slightly. Although some details may be lost during the ELM spatial enhancement, the low RMSEs and the strong CCs indicate that the ELM can preserve much of the physical TIR information in the original input TIR images for quantitative remote sensing applications.

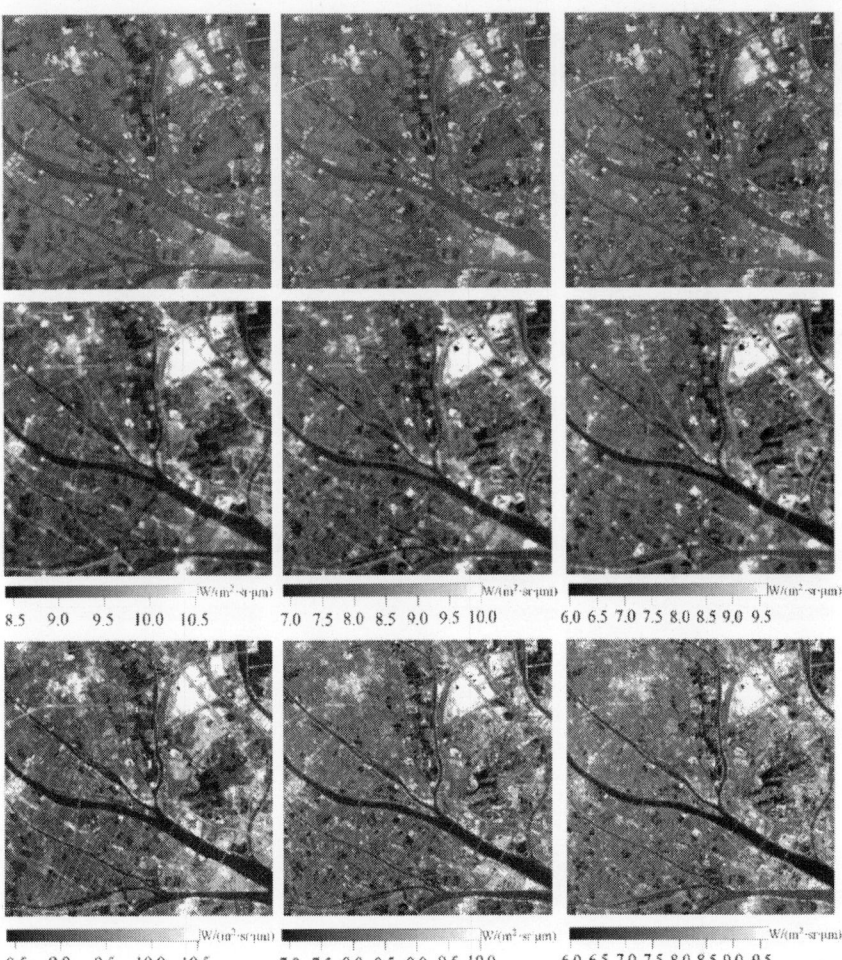

Figure 5. Thermal spatial sharpening results of testing experiments via ELM algorithm. False colour images of downscaled Landsat ETM+ multispectral data at 60 m (**upper row**), downscaled TIR images at 120 m (**middle row**) and sharpened TIR images at 60 m (**lower row**). The original TIR images at 60 m refer to the middle row in Figure 6. From left to right, they were acquired on 20 October, 7 December and 23 December 2013.

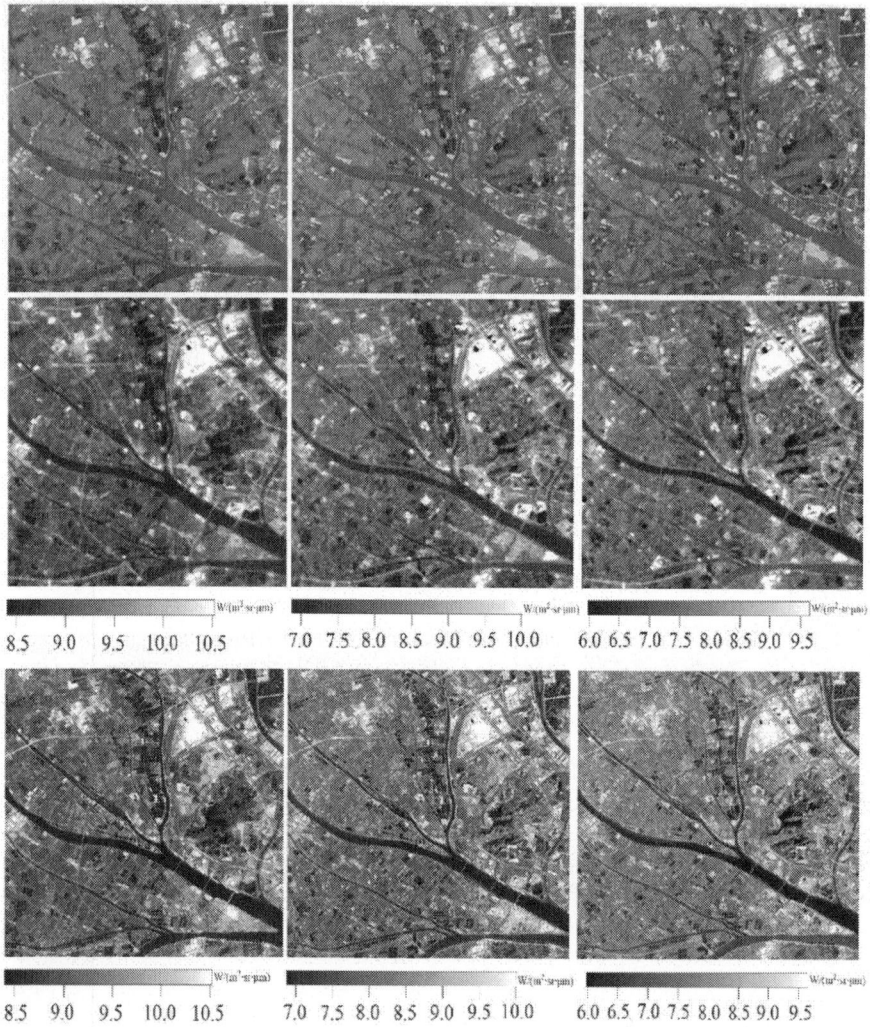

8.5 9.0 9.5 10.0 10.5   7.0 7.5 8.0 8.5 9.0 9.5 10.0   6.0 6.5 7.0 7.5 8.0 8.5 9.0 9.5

8.5 9.0 9.5 10.0 10.5   7.0 7.5 8.0 8.5 9.0 9.5 10.0   6.0 6.5 7.0 7.5 8.0 8.5 9.0 9.5

Figure 6. Thermal spatial sharpening results of simulated experiments using ELM algorithm. False colour images of observed Landsat ETM+ multispectral data at 30 m (**upper row**), TIR images at 60 m (**middle row**) and Enhanced TIR images at 30 m (**lower row**). From left to right, they were acquired on 20 October, 7 December and 23 December 2013.

Table 1. Quantitative assessment of thermal sharpening in the testing experiment via ELM.

Date	20 October 2013	7 December 2013	23 December 2013
CC	0.8788 *	0.8251 *	0.8017 *
RMSE	0.0844	0.0891	0.0909
AD	−9.8940e-06	−2.1980e-06	6.3913e-06
AAD	0.0595	0.0620	0.0614

In the second step of the proposed fusion model, considering the characteristics of complex surface in this study area, the size of the searching window was set to three MODIS pixels and there are five land cover types in the SADFAT. To ensure the accuracy of searching similar pixels and obtain the NDVI values at the prediction date for LST retrieval, the reflectance values of both Band 3 and Band 4 combined with the thermal radiance information from Band 6 of Landsat ETM+ and its corresponding bands of MODIS at t_1 and t_2 were input into the SADFAT fusion model developed by Weng *et al.* [42].Figure 7 and Figure 8 showed the results of predicted images at 60 m and 30 m on 7 December 2013, respectively. For both the testing and simulated experiments, the predicted LST images (Figure 7b and Figure 8b) are visually similar to the original Landsat LST images (Figure 7c and Figure 8c), but they provide higher spatial details in terms of thermal information.

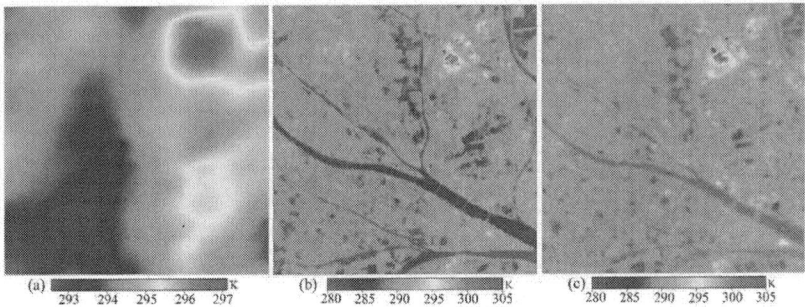

Figure 7. SADFAT result in the testing experiment. **(a)** MODIS LST images resampled from 1 km to 60 m; **(b)** SADFAT-derived image at 60 m; **(c)** original Landsat ETM+ LST image at 60 m spatial on 7 December 2013.

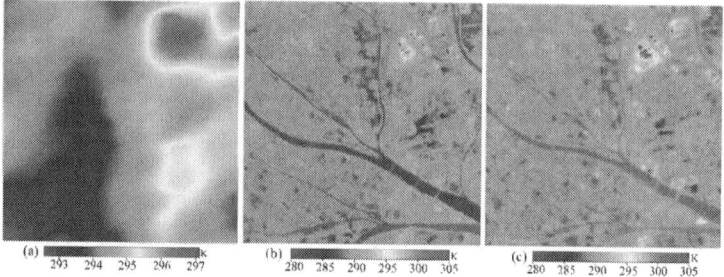

Figure 8. SADFAT result in simulated experiment. **(a)** MODIS LST images resampled from 1 km to 30 m; **(b)** SADFAT-derived image at 30 m; **(c)** up-scaled Landsat ETM+ LST image at 30 m spatial on 7 December 2013.

Figure 9 displays the scatter plots between the predicted and observed LSTs for testing experiment on 7 December 2013. It shows a strong agreement between predicted and observed LSTs in testing experiment. Nevertheless, there are slight differences in some pixels between the predicted and observed LSTs, owing to the limitation of SADFAT that may not be suitable for the changes in land cover and other surface conditions during the prediction period [42]. Table 2 shows the results of CC, RMSE, AD and AAD between the predicted and observed LSTs for the testing experiment at 60 m. The values of AD and AAD are smaller than 2.0 K whereas the CC values are larger than 0.75 suggesting significant consistency.

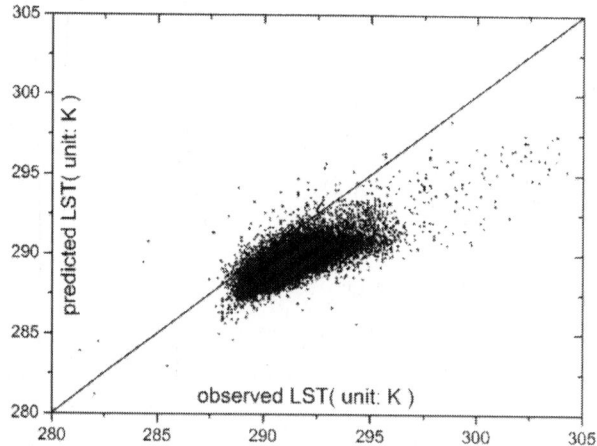

Figure 9. Scatter plots between the predicted and original LSTs at 60 m of the testing experiment on 7 December 2013. x-axis denotes the original LSTs, and y-axis denotes the prediction LSTs.

Table 2. Indicators of prediction accuracy on 7 December 2013 (unit: K).

Indicator	CC	RMSE	AD	AAD
testing experiment	0.7554 *	1.8242	1.6354	1.6498

DISCUSSION

The proposed fusion model employs ELM to enhance the spatial resolution of TIR data before using the spatial-temporal fusion algorithm. As shown in Figure 5 and 6, the sharpened images of both testing and simulated experiments have more spatial details compared with original Landsat TIR images. As shown in Figure 7b and Figure 8b, it can be observed that the predicted LST images provide higher spatial details in terms of thermal information than the original Landsat LST images (Figure 7c andFigure 8c) in both the testing and simulated experiments. In Figure 9, the scatter plot of the testing experiment shows a strong relationship between predicted and original LSTs at 60 m. In addition, small values of AD and AAD (Table 2) (less than 1.7 K) and high values of CC (larger than 0.75) are observed. These results show that our proposed method can enhance the spatial resolution of the predicted LST image and also preserve the spectral information simultaneously.

However, there are still some limitations/errors in the proposed method, including: (i) the thermal spatial sharpening ELM algorithm is highly time-consuming; (ii) some important information in the original TIR images may be lost during the ELM spatial enhancement; (iii) the proposed fusion method cannot predict the changes in LST that are not presented on the MODIS and/or Landsat images, especially in the missing pixels caused by cloud contamination [42]; (iv) the proposed fusion method requires at least two input pairs of fine and coarse resolution images, in the same season or under similar atmospheric conditions; (v) the number of land cover types and the size of a moving window should be determined by searching similar pixels in SADFAT, which may limit the automated process. Thus, more research on the following issues will be conducted in the near future: (i) by improving the fusion accuracy and efficiency of ELM simultaneously, and (ii) developing a more advanced fusion method for generating synthetic LST images to resolve the limitations of seasonal changes and different atmospheric conditions.

CONCLUSION

In this paper, a novel data fusion method by integrating a thermal spatial sharpening algorithm with spatial-temporal fusion to generate synthetic LST datasets at both high spatial and temporal revolution was developed. The performance of this method was tested using Landsat ETM+ and MODIS images both in testing and simulated experiments. At first, the extreme learning machine algorithm was selected to enhance the spatial resolution of the Landsat ETM+ TIR data. Then, the MODIS LST and enhanced Landsat ETM+ TIR data were fused using SADFAT developed by Weng *et al.* [42] to derive high temporal resolution synthetic data. The proposed image fusion method provides an alternative to generate the synthetic high resolution image for remote sensing applications from multi-source satellite data. Compared with the traditional spatio-temporal adaptive data fusion algorithm, the synthetic LST images derived in this study can depict more spatial details. The generated synthetic LST product can be used for monitoring the variation of land surface temperature in urban heat island studies. Further research on other sensors such as Landsat TIRS, HJ-1B satellite will be conducted.

ACKNOWLEDGMENTS

This research was sponsored by the grant PolyU 1-ZVBR from the Research Institute for Sustainable Urban Development, the Hong Kong Polytechnic University, grant G-YM85 from the Hong Kong Polytechnic University, and grant F-PP1Q from the Early Career Scheme of the Hong Kong Research Grants Council. This work was also jointly supported by the National Basic Research Program (973 Program, Grant No.2011CB707102), and the project funded by the Priority Academic Program Development (PAPD) of Jiangsu Higher Education Institutions. The authors wish to thank Prof. Qihao Weng, Dr. Feng Gao, and Dr. Xiaolin Zhu for their valuable suggestions in this study; We would also like to thank the NASA Goddard Space Flight Centre and U.S. Geological Survey for the MODIS and Landsat images.

AUTHOR CONTRIBUTIONS

Yai Bai, Man Sing Wong and Kai Qin conceived and designed the experiments, performed the experiments and analyzed the data; Yai Bai, Man Sing Wong, Wen-zhong Shi, Li-xin Wu and Kai Qin prepared the manuscript.

APPENDIX A. EXTREME LEARNING MACHINE (ELM) ALGORITHM

Extreme Learning Machine (ELM) proposed by Huang *et al.* [45] is designed for single-hidden layer feed-forward neural networks (SLFNs) that can adjust the input weights and determine the output weights analytically. Supposing the number of training samples $(x_i,\ t_i)$ is N, where xi=[xi1,xi2,$\cdots$,xin] T $\in$ Rn and t_i=[t_{i1}, t_{i2},$\cdots$,t_{im}] T $\in$ R$_m$, the standard SLFNs equipped with N$^\sim$ hidden neurons and activation function g(x)=sig(w_i·x_i+b_i) can be mathematically expressed as:

$$H\beta = T \tag{7}$$

where

$$H\left(\mathbf{w}_1, \cdots, \mathbf{w}_{\tilde{N}}, b_1, \cdots, b_{\tilde{N}}, \mathbf{x}_1, \cdots, \mathbf{x}_{\tilde{N}}\right) =$$

$$\begin{bmatrix} g\left(\mathbf{w}_1 \cdot \mathbf{x}_1 + b_1\right) & \cdots & g\left(\mathbf{w}_{\tilde{N}} \cdot \mathbf{x}_1 + b_{\tilde{N}}\right) \\ \vdots & \cdots & \vdots \\ g\left(\mathbf{w}_1 \cdot \mathbf{x}_N + b_1\right) & \cdots & g\left(\mathbf{w}_{\tilde{N}} \cdot \mathbf{x}_N + b_{\tilde{N}}\right) \end{bmatrix}_{N \times \tilde{N}}$$

$$\beta = \begin{bmatrix} \beta_1^T \\ \vdots \\ \beta_{\tilde{N}}^T \end{bmatrix}_{\tilde{N} \times m} \quad \text{and} \quad T = \begin{bmatrix} t_1^T \\ \vdots \\ t_N^T \end{bmatrix}_{N \times m},$$

And H represents the hidden layer output matrix of the neural network; bi is the threshold of the i_{th} hidden neuron; w_i=[w_{i1}, w_{i2},$\cdots$,w_{in}]T denotes the weight vector connecting the i_{th} hidden neuron and the input neurons; $\beta_i = [\beta_{i1}, \beta_{i2}, L, \beta_{im}]T$ stands for the weight vector connecting the i_{th} hidden neuron and the output neurons.

APPENDIX B. SPATIO-TEMPORAL ADAPTIVE DATA FUSION ALGORITHM FOR TEMPERATURE MAPPING (SADFAT)

The Spatio-Temporal Adaptive Data Fusion Algorithm for Temperature Mapping (SADFAT), developed by Weng *et al.* [42], was used in this study to predict synthetic thermal radiance and LST images in the second step. Below is a brief description of the algorithm:

If there are two pairs of Landsat ETM+ and MODIS image acquired at t_1 and t_2, respectively, the changes of radiance of a M pixel between t_1 and t_2 can be computed as:

$$R_M(t_2) - R_M(t_1) = \frac{1}{k_1} \sum_{i=1}^{N} l_i (R_{iL}(t_2) - R_{iL}(t_1))$$

(8)

where R denotes the radiance at the satellite sensor, N is the pixel number of Landsat within a MODIS pixel and l_i denotes the fraction of each Landsat pixel, t is the acquisition date, and k_1 is the coefficient for the relative adjustment for the Landsat and MODIS radiance pixels. Considering the seasonal change of LST based on ATC, and through the Planck's law, the radiance change of an L pixel from t_1 to t_2 can be expressed as:

$$R_{iL}(t_2) - R_{iL}(t_1) = 2d \cos\left(\omega_i + a\frac{e_1 + e_2}{2}\right) \sin\left(a\frac{e_2 - e_1}{2}\right) = D \cos(\omega_i + a\bar{e})$$

(9)

where a is the angular frequency, ω is the phase shift, or heat lag, d is the amplitude of the radiance variation, D is the constant, and $\bar{e}$ is the mean acquisition date, e_1 and e_2 are the parameters input to the algorithm. Therefore, if the radiances of the j_{th} L pixel at date t_1 and t_2 are known, the equation can be formulated as Equation (10):

$$R_{jL}(t_2) - R_{jL}(t_1) = 2d \cos(\omega_i + a\bar{e}) \sin\left(a \frac{e_2 - e_1}{2}\right)$$

$$(10)$$

By combining Equation (9) with Equation (10), the Equation (11) indicates that the ratio of the radiance change of j_{th} L pixel to the radiance of the corresponding M pixel is constant for a certain L pixel, which can be quantified as below:

$$\frac{R_{jL}(t_2) - R_{jL}(t_1)}{R_M(t_2) - R_M(t_1)} = \frac{\cos(\omega_k + a\bar{e})}{\frac{1}{k_1} \sum_{i=1}^{N} l_i \cos(\omega_i + a\bar{e})} = h_j$$

$$(11)$$

where hj is named the conversion coefficient for the purpose of consistency, ω reflects the phase shift of a pixel and is related with the thermal characteristics of land cover.

Supposing s is the moving window size, if there is a pair of L and M radiance images at t0 and a M radiance image at tp, the predicted L radiance image at tp can be predicted by the following equation:

$$R_L(x_{s/2}, y_{s/2}, t_p) = R_L(x_{s/2}, y_{s/2}, t_0) + \sum_{i=1}^{N} W_i \times h_i \times [R_M(x_i, y_i, t_p) - R_M(x_i, y_i, t_0)]$$

$$(12)$$

where W_i represents the weight of a neighbouring similar pixel, and N is the number of the spectral similar pixel. In this paper, after predicting the radiance image using SADFAT method (Weng *et al.* [42]), enhanced Landsat ETM+ TIR data with both high spatial and temporal resolution would be converted to LST using the generalized single channel method [46]. The following equations are the implementation procedures of the generalized single channel method for ETM+ TIR data.

$$T_{surface} = \alpha \left[\varepsilon^{-1} \left(\varphi_1 R_{sensor} + \varphi_2 \right) + \varphi_3 \right] + \beta$$

(13)

$$\alpha = \left(\frac{c_2 R_{sensor}}{T_{sensor}^2} \left(\frac{\lambda^4}{c_1} R_{sensor} + \lambda^{-1} \right) \right)^{-1} \text{ and } \beta = -\alpha R_{sensor} + T_{sensor}$$

(14)

where $T_{surface}$ is the retrieval temperature of the surface, R_{sensor} and T_{sensor} stands for the radiance and brightness temperature at-satellite, respectively, λ (11.3355 μm) is the effective wavelength for ETM+ sensor, c1, c2 are the constants, ε is the land surface emissivity. The atmospheric functions are described with the water vapour content:

$$\begin{bmatrix} \varphi_1 \\ \varphi_2 \\ \varphi_3 \end{bmatrix} = \begin{bmatrix} 0.14714 & -0.15583 & 1.1234 \\ -1.1836 & -0.37607 & -0.52894 \\ -0.04554 & 1.8719 & -0.39071 \end{bmatrix} \begin{bmatrix} v^2 \\ v \\ 1 \end{bmatrix}$$

(15)

where v is the water vapour content, which can be obtained from the water vapour content product (MOD05_L2).

REFERENCES

1.   Kalma, J.D.; McVicar, T.R.; McCabe, M.F. Estimating land surface evaporation: A review of methods using remotely sensed surface temperature data. *Surveys Geophys.* 2008, *29*, 421–469.
2.   Cammalleri, C.; Anderson, M.; Ciraolo, G.; D'Urso, G.; Kustas, W.; La Loggia, G.; Minacapilli, M. Applications of a remote sensing-based two-source energy balance algorithm for mapping surface fluxes without *in situ* air temperature observations. *Remote Sens. Environ.* 2012, *124*, 502–515.
3.   Srivastava, P.K.; Han, D.; Ramirez, M.R.; Islam, T. Machine learning techniques for downscaling smos satellite soil moisture using modis land

surface temperature for hydrological application. *Water Resour. Manag.* 2013, *27*, 3127–3144.

4.   Song, X.; Leng, P.; Li, X.; Li, X.; Ma, J. Retrieval of daily evolution of soil moisture from satellite-derived land surface temperature and net surface shortwave radiation. *Int. J. Remote Sens.* 2013, *34*, 3289–3298.

5.   Bateni, S.; Entekhabi, D.; Castelli, F. Mapping evaporation and estimation of surface control of evaporation using remotely sensed land surface temperature from a constellation of satellites. *Water Resour. Res.* 2013, *49*, 950–968.

6.   Tang, R.; Li, Z.-L.; Jia, Y.; Li, C.; Chen, K.-S.; Sun, X.; Lou, J. Evaluating one-and two-source energy balance models in estimating surface evapotranspiration from Landsat-derived surface temperature and field measurements. *Int. J. Remote Sens.* 2013, *34*, 3299–3313.

7.   Anderson, M.; Kustas, W.; Norman, J.; Hain, C.; Mecikalski, J.; Schultz, L.; González-Dugo, M.; Cammalleri, C.; D'Urso, G.; Pimstein, A. Mapping daily evapotranspiration at field to continental scales using geostationary and polar orbiting satellite imagery. *Hydrol. Earth Syst. Sci.* 2011, *15*, 223–239.

8.   Wong, M.S.; Nichol, J.E. Spatial variability of frontal area index and its relationship with urban heat island intensity. *Int. J. Remote Sens.* 2013, *34*, 885–896.

9.   Weng, Q.; Fu, P. Modeling annual parameters of clear-sky land surface temperature variations and evaluating the impact of cloud cover using time series of landsat tir data. *Remote Sens. Environ.* 2014, *140*, 267–278.

10. Goward, S.N.; Masek, J.G.; Williams, D.L.; Irons, J.R.; Thompson, R. The landsat 7 mission: Terrestrial research and applications for the 21st century. *Remote Sens. Environ.* 2001, *78*, 3–12.

11. Anderson, M.C.; Allen, R.G.; Morse, A.; Kustas, W.P. Use of landsat thermal imagery in monitoring evapotranspiration and managing water resources. *Remote Sens. Environ.* 2012, *122*, 50–65.

12. Almeida, T.; de Souza Filho, C.; Rossetto, R. Aster and landsat ETM+ images applied to sugarcane yield forecast. *Int. J. Remote Sens.* 2006, *27*, 4057–4069.

13. Sameen, M.I.; Al Kubaisy, M.A. Automatic surface temperature mapping in arcgis using landsat-8 tirs and envi tools, case study: Al Habbaniyah Lake. *J. Environ. Earth Sci.* 2014, *4*, 12–17.

14. Li, Y.-Y.; Zhang, H.; Kainz, W. Monitoring patterns of urban heat islands of the fast-growing shanghai metropolis, china: Using time-series of landsat tm/etm+ data. *Int. J. Appl. Earth Obs. Geoinf.* 2012, *19*, 127–138.

15. Masek, J.G.; Collatz, G.J. Estimating forest carbon fluxes in a disturbed southeastern landscape: Integration of remote sensing, forest inventory, and biogeochemical modeling. *J. Geophys. Res.*2006, *111*.

16. Ju, J.; Roy, D.P. The availability of cloud-free landsat ETM+ data over the conterminous united states and globally. *Remote Sens. Environ.* 2008, *112*, 1196–1211.

17. Leckie, D.G. Advances in remote sensing technologies for forest surveys and management. *Can. J. For. Res.* 1990, *20*, 464–483.

18. Justice, C.O.; Vermote, E.; Townshend, J.R.; Defries, R.; Roy, D.P.; Hall, D.K.; Salomonson, V.V.; Privette, J.L.; Riggs, G.; Strahler, A. The moderate resolution imaging spectroradiometer (modis): Land remote sensing for global change research. *IEEE Trans. Geosci. Remote Sens.*1998, *36*, 1228–1249.

19. Stathopoulou, M.; Cartalis, C. Downscaling avhrr land surface temperatures for improved surface urban heat island intensity estimation. *Remote Sens. Environ.* 2009, *113*, 2592–2605.

20. Pohl, C.; Van Genderen, J. Review article multisensor image fusion in remote sensing: Concepts, methods and applications. *Int. J. Remote Sens.* 1998, *19*, 823–854.

21. Smith, M.I.; Heather, J.P. A review of image fusion technology in 2005. In *Defense and Security, 2005*; International Society for Optics and Photonics: Bellingham WA, USA, 2005; pp. 29–45.

22. Ha, W.; Gowda, P.H.; Howell, T.A. A review of potential image fusion methods for remote sensing-based irrigation management: Part II. *Irrig. Sci.* 2013, *31*, 851–869.

23. González-Audícana, M.; Saleta, J.L.; Catalán, R.G.; García, R. Fusion of multispectral and panchromatic images using improved IHS and PCA mergers based on wavelet decomposition. *IEEE Trans. Geosci. Remote Sens.* 2004, *42*, 1291–1299.

24. Naidu, V.; Raol, J. Pixel-level image fusion using wavelets and principal component analysis. *Def. Sci. J.* 2008, *58*, 338–352.

25. Tu, T.-M.; Huang, P.S.; Hung, C.-L.; Chang, C.-P. A fast intensity-hue-saturation fusion technique with spectral adjustment for ikonos imagery. *IEEE Geosci. Remote Sens. Lett.* 2004, *1*, 309–312.

26. Choi, M. A new intensity-hue-saturation fusion approach to image fusion with a tradeoff parameter. *IEEE Trans. Geosci. Remote Sens.* 2006, *44*, 1672–1682.

27. Amolins, K.; Zhang, Y.; Dare, P. Wavelet based image fusion techniques—An introduction, review and comparison. *ISPRS J. Photogramm. Remote Sens.* 2007, *62*, 249–263.

28. Zhang, Y.; Hong, G. An ihs and wavelet integrated approach to improve pan-sharpening visual quality of natural colour ikonos and quickbird images. *Inf. Fus.* 2005, *6*, 225–234.

29. Zhan, W.; Chen, Y.; Zhou, J.; Li, J.; Liu, W. Sharpening thermal imageries: A generalized theoretical framework from an assimilation perspective. *IEEE Trans. Geosci. Remote Sens.* 2011,*49*, 773–789.

30. Rodriguez-Galiano, V.; Pardo-Iguzquiza, E.; Sanchez-Castillo, M.; Chica-Olmo, M.; Chica-Rivas, M. Downscaling Landsat 7 ETM+ thermal imagery using land surface temperature and NDVI images. *Int. J. Appl. Earth Obs. Geoinf.* 2012, *18*, 515–527.

31. Rodriguez-Galiano, V.; Ghimire, B.; Pardo-Igúzquiza, E.; Chica-Olmo, M.; Congalton, R. Incorporating the downscaled landsat tm thermal band in land-cover classification using random forest. *Photogramm. Eng. Remote Sens.* 2012, *78*, 129–137.

32. Hong, S.-H.; Hendrickx, J.M.; Borchers, B. Up-scaling of sebal derived evapotranspiration maps from Landsat (30m) to MODIS (250m) scale. *J. Hydrol.* 2009, *370*, 122–138.

33. Jeganathan, C.; Hamm, N.; Mukherjee, S.; Atkinson, P.M.; Raju, P.; Dadhwal, V. Evaluating a thermal image sharpening model over a mixed agricultural landscape in India. *Int. J. Appl. Earth Obs. Geoinf.* 2011, *13*, 178–191.

34. Gao, F.; Masek, J.; Schwaller, M.; Hall, F. On the blending of the Landsat and MODIS surface reflectance: Predicting daily landsat surface reflectance. *IEEE Trans. Geosci. Remote Sens.* 2006,*44*, 2207–2218.

35. Zhu, X.; Chen, J.; Gao, F.; Chen, X.; Masek, J.G. An enhanced spatial and temporal adaptive reflectance fusion model for complex heterogeneous regions. *Remote Sens. Environ.* 2010, *114*, 2610–2623.

36. Hilker, T.; Wulder, M.A.; Coops, N.C.; Seitz, N.; White, J.C.; Gao, F.; Masek, J.G.; Stenhouse, G. Generation of dense time series synthetic landsat data through data blending with MODIS using a spatial and temporal adaptive reflectance fusion model. *Remote Sens. Environ.* 2009, *113*, 1988–1999.

37. Roy, D.P.; Ju, J.; Lewis, P.; Schaaf, C.; Gao, F.; Hansen, M.; Lindquist, E. Multi-temporal modis—Landsat data fusion for relative radiometric normalization, gap filling, and prediction of landsat data. *Remote Sens. Environ.* 2008, *112*, 3112–3130.

38. Zurita-Milla, R.; Clevers, J.G.; Schaepman, M.E. Unmixing-based Landsat TM and MERIS FR data fusion. *IEEE Geosci. Remote Sens. Lett.* 2008, *5*, 453–457.

39. Huang, B.; Song, H. Spatiotemporal reflectance fusion via sparse representation. *IEEE Trans. Geosci. Remote Sens.* 2012, *50*, 3707–3716.

40. Song, H.; Huang, B. Spatiotemporal satellite image fusion through one-pair image learning. *IEEE Trans. Geosci. Remote Sens.* 2013, *51*, 1883–1896.

41. Huang, B.; Wang, J.; Song, H.; Fu, D.; Wong, K. Generating high spatiotemporal resolution land surface temperature for urban heat island monitoring. *IEEE Geosci. Remote Sens. Lett.* 2013, *10*, 1011–1015.

42. Weng, Q.; Fu, P.; Gao, F. Generating daily land surface temperature at landsat resolution by fusing Landsat and MODIS data. *Remote Sens. Environ.* 2014, *145*, 55–67.

43. Huang, G.-B.; Zhou, H.; Ding, X.; Zhang, R. Extreme learning machine for regression and multiclass classification. *IEEE Trans. Syst. Man. Cybern. B Cybern.* 2012, *42*, 513–529. [PubMed]

44. Yao, W.; Han, M. Fusion of thermal infrared and multispectral remote sensing images via neural network regression. *J. Image Gr.* 2010, *15*, 1278–1284. [Google Scholar]

45. Huang, G.-B.; Zhu, Q.-Y.; Siew, C.-K. Extreme learning machine: A new learning scheme of feedforward neural networks. In Proceedings of the 2004 IEEE International Joint Conference on Neural Networks, Budapest, Hungary, 25–29 July 2004; pp. 985–990.

46. Jiménez-Muñoz, J.C.; Sobrino, J.A. A generalized single-channel method for retrieving land surface temperature from remote sensing data. *J. Geophys. Res.* 2003, *108*.

47. Wan, Z.; Zhang, Y.; Zhang, Q.; Li, Z.-L. Quality assessment and validation of the MODIS global land surface temperature. *Int. J. Remote Sens.* 2004, *25*, 261–274.

48.

CITATION

Yang Bai, Man Sing Wong, Wen-Zhong Shi, Li-Xin Wu and Kai Qin, Advancing of Land Surface Temperature Retrieval Using Extreme Learning Machine and Spatio-Temporal Adaptive Data Fusion Algorithm, doi:10.3390/rs70404424

CHAPTER 4

GIS Applied to Integrated Coastal Zone and Ocean Management: Mapping, Change Detection and Spatial Modeling for Coastal Management in Southern Brazil

Tatiana S. da Silva, Maria Luiza Rosa and Flávia Farina

Federal University of Rio Grande do Sul, Institute of Geoscience, Brazil

INTRODUCTION

Information is the basis for sustainable development. If a decision is taken without any quality information to back it up, it relies on guesswork. Accurate, comprehensive and periodic environmental information is then crucial for the success of the decision-making and environmental planning processes.

Coastal zones are characterized by fragile, complex and productive environments, typical of the sea-land system. They deserve special attention from the government and society. Besides, most of the world population lives by the sea and there is a permanent trend of demographic concentration in these areas. The health and well-being of the coastal populations, and sometimes even their survival, depend on the status of coastal and marine ecosystems. Managing this complexity implies the cooperation between the levels of the government and the society.

In Brazil, the National Coastal Management Plan (PNGC) was established through the Law 7661 of 1988 in order to plan the use of coastal areas. Today in its second version, the PNGC provides the following instruments: (1) the State Coastal Management Plan (PEGC), clarifies the development of the PNGC at the state level, aiming to implement a State Coastal Management Policy; (2) the Municipal Coastal Management Plan (PMGC), clarifies the development of the PNGC and PEGC at the municipal level, aiming to implement a Municipal Coastal Management Policy; (3) the Coastal Management Information System (SIGERCO) is a component of the National Environmental Information System (SINIMA) and gives support to the state / municipal subsystems; (4) the Environmental Monitoring System (SMA-ZC) is the operational framework of continuous data collection to monitor the social-environmental indicators and support the management plans in the coastal zone; (5) the Environmental Quality Report (RQA-ZC) is the consolidation of results from the SMA-ZC in periodic reports and, above all, aims to assess the efficiency of the management actions undertaken; (6) the Ecologic-Economic Coastal Zoning (ZEEC) is the spatial regulatory instrument to plan land use in a given territory; and (7) the Coastal Zone Management Plan (PGZC) is a set of coordinated and programmatic actions, built in a participatory manner, and applicable to different levels of government.

Unfortunately, more than 20 years after the implementation of the 7661 Law, the institutionalization of coastal management is still incipient (Jablonski and Filet, 2008), even in the Rio Grande do Sul State, where the environmental control is very restrictive compared to the rest of the country. In some municipalities of the Rio Grande do Sul coastal plain, environmental plans were built and consist in the only environmental regulatory instruments at municipal level, but they are not truly implemented. At the national level, on the other hand, the adoption of the I3Geo as component of SINIMA was a successful initiative. The I3Geo is a platform for publishing spatial data and interactive mapping applications to the web, helping the establishment of cooperative networks, showing the advantages of using a GIS platform as the core of an environmental information system.

In a general sense, all the coastal management instruments, as they are designed in the PNGC, explicitly depend on spatial tools or at least would be benefited through the use of maps. The universities focused on the coastal and marine ecosystems of the Rio Grande do Sul generate a number of GIS-based products as research results. They have been helpful in building the environmental and master plans of many coastal municipalities, but there is still a great potential to include spatial information in other mechanisms of coastal management. A significant knowledge about the natural resources in this region has been gained from GIS and remote sensing in the last 30 years. Geotechnology has been successfully used to understand the spatial structure and dynamics of the coastal landscapes and, more recently, in simulation modeling and change prediction. The acquired knowledge should be included into the existing instruments, and also used to improve them, promoting the adaptation to the current pace and reach of human activities over the coastal zone.

The Rio Grande do Sul coastal zone (figure 1), in Southern Brazil, is characterized by a wide coastal plain generated by the sea level changes during the Quaternary, which resulted in a complex lagoon system all over its extent (Asmus *et al.*, 1988). The Patos Lagoon is the most expressive of such water bodies, comprising almost 10.000 km² and a number of valuable marginal ecosystems and aquatic species, some of them of economic interest. Research has been focused on coastal issues for more than 20 years. And since then, space has been the base of approach.

Thus, this chapter aims to provide a discussion about the applicability of the mapping, change detection and spatial modeling efforts in the Southern Brazil coastal plain to the policy instruments defined by the National Coastal Management Plan. We also recommend strategies to promote an adaptative management concerning coastal zones through the use of GIS. Geotechnology will be presented as a way to enhance the exchange and feedback among academic researchers, stakeholders and community.

Figure 1. Rio Grande do Sul coastal zone.

THE BRAZILIAN PROGRAM ON COASTAL MANAGEMENT: IMPLEMENTATION STATUS AND SPATIAL NATURE OF POLICY INSTRUMENTS

The second National Plan of Coastal Management – PNGC II was approved in 1997, in substitution of PNGC I. The National Program on Coastal and Ocean Management (GERCOM) aims to put PNGC into operation, in order to plan and manage the economic activities so to ensure the sustainable use of the coastal environments. GERCO is coordinated by the Ministry of Environment and executed by the 17 coastal states. Rio Grande do Sul is one of the most advanced states regarding the implementation of the coastal management plan, but actions are highly concentrated in the north littoral where urbanization is more spread and severe.

The State Envionmental Protection Agency (FEPAM) is the state-level authority in charge of environmental management in Rio Grande do Sul. FEPAM is legally in charge of implementing the Coastal Management Program. The participation of FEPAM, as the ultimate client for the management plans, is crucial to guarantee long-term sustainability of applied research endeavors. To the extent that FEPAM takes ownership of projects results, and makes the proposed plans its own, this would guarantee that such plans are incorporated in the State's budged and overall environmental policy (Tagliani *et al.*, 2003).

The Rio Grande do Sul coastal zone comprises 27 coastal municipalities: Torres, Arroio do Sal, Três Cachoeiras, Três Forquilhas, Maquiné, Capão da Canoa, Terra da Areia, Xangrilá, Osório, Imbé, Tramandaí, Cidreira, Palmares do Sul, Viamão, Mostardas, Barra do Ribeiro, Tapes, Tavares, Camaquã, Arambé, São José do Norte, São Lourenço do Sul, Rio Grande, Pelotas, Arroio Grande, Jaguarão, and Santa Vitória do Palmar. Only 9 of them have environmental plans approved by FEPAM. However, conservation units and preservation areas are considered in most of the master plans. In some municipalities, they are the only instrument of environmental and territorial control.

The State Plan of Coastal Management (PEGC) and the Municipal Plan of Coastal Management (PMGC) are highly related to territorial ordination planning. Environmental zoning proposals as well as the detection of priority areas for management subside state and municipal plans. This is particularly important for those municipalities where the lack of human and technological resources is more dramatic.

Environmental zoning proposals can also base the Coastal Ecological-Economic Zoning (ZEEC), which aims to regulate the territorial use in order to achieve environmental sustainability of the coastal zone development, respecting the directives of the Ecological-Economic Zoning in national scale.

GIS-based project results in general are potential inputs for the Coastal Management Information System (SIGERCO), a component of the National System of Environmental Information (SINIMA). SIGERCO integrates PNGC information, aiming to give support and capillarity to the subsystems structured and managed by coastal states and municipalities.

Research results that also include a temporal dimension (land use change studies, for example) give GIS procedures the status of a monitoring method, once they can be replicated in other times or locations. This type of spatial data can be used as environmental quality indicators and, thus, can be incorporated by the Coastal Zone Environmental Monitoring System (SMA-ZC). Consequently, they are able to support the Coastal Zone Environmental Quality Report (RQA-ZC), which is the periodic consolidation of the results obtained by the environmental monitoring.

Once the institutions involved in public management have absorbed the available technology and information, the possibilities of use them in the decision making process are diverse. The possibility to visualize any attribute in a map makes easier to understand the coastal space, promoting knowledge-based public participation in the decision making process. The chance of governance to succeed is higher in this way.

In a general sense, all the coastal management instruments, as they are designed in the PNGC, explicitly depend on spatial tools or at least would be benefited through the use of maps. Some of the GIS-based research works have already been helpful in building the environmental and master plans of some coastal municipalities (Pelotas, Rio Grande, São Lourenço do Sul, Turuçu, among others) but there is still a great potential to include this information in other mechanisms of coastal management. Geotechnology has been successfully used to understand the spatial structure and dynamics of medium littoral of the Rio Grande do Sul coastal plain. The acquired knowledge should also be included into the existing instruments to improve them, providing a more prospective focus and promoting the adaptation to the current pace and reach of human activities over the coastal zone.

RIO GRANDE DO SUL COASTAL PLAIN: MAPPING, CHANGE DETECTION AND SPATIAL MODELING AS DECISION SUPPORT RESOURCES

Mapping the Coastal Zone

PNGC delegates power for states and municipalities to legislate issues related to the use of soil, water and forest resources. Based on that, municipalities are highly recommended to create master plans and legal instrumentsto manage land use, mantainenvironmental quality, and use properly the natural resources at municipal level. Several methods are involved in the implementation of a master plan, like statistical analysis, mapping, zoning, registration survey, field research, among others. Thus, GIS represents an extremely useful tool for municipal planning purposes. GIS gathers together a great set of application to collect, storage, restore, change, and represent spatial data, as well as related attributes. GIS implementation, as the basis for planning and management, means a huge step toward a greater efficiency of municipal administration.

Basic GIS questions, such as *when, what, what distance, and which*, sometimes need integrating tools to be answered. In this sense, Multi Criteria Evaluation (MCE) techniques are effective procedures for several purposes. Gómez & Barredo (2005) have analyzed GIS analytical functions integrated to MCE techniques, which allows introducing objective and/or heuristic knowledge to simulate the possible results of different decisions and develop virtual scenarios to evaluate the implementation of policies.

Geological and Geomorphologic Mapping of Coastal Plain as the Basis for Land Use Planning

The knowledge about the substrate and geological evolution is a very important point for planning the use of coastal environments. When considering the substrate characteristics, we might evaluate the best way and locationfor development, integrating human interests with the particularities of each environment.

Rio Grande do Sul coastal plain geology has been studied over the years especially through surface geological mapping, supported by remote sensing data, drilling, and more recently, geophysical surveys. This information can be integrated and made available through GIS, allowing its use in several applications.

Rio Grande do Sul coastal plain represents the top and youngest portion of the Pelotas sedimentary basin. It is mainly formed by alluvial fan and barrier-lagoon sedimentary deposits (figure 2). According to previous works, these deposits were formed in response to sea level changes, which were controlled by glacioeustasy during the Quaternary. Mineralogical and geomorphological patterns result in four barrier-lagoon systems. The oldest system is designated as I and the youngest (still active) as IV (Tomazelli & Villwock, 1996). The age of the oldest systems was established mainly by correlation with oxygen isotope curves. Each system represents the maximum of the Postglacial Marine Transgression (PMT): 400 ka (I), 325 ka (II) and 120 ka (III).

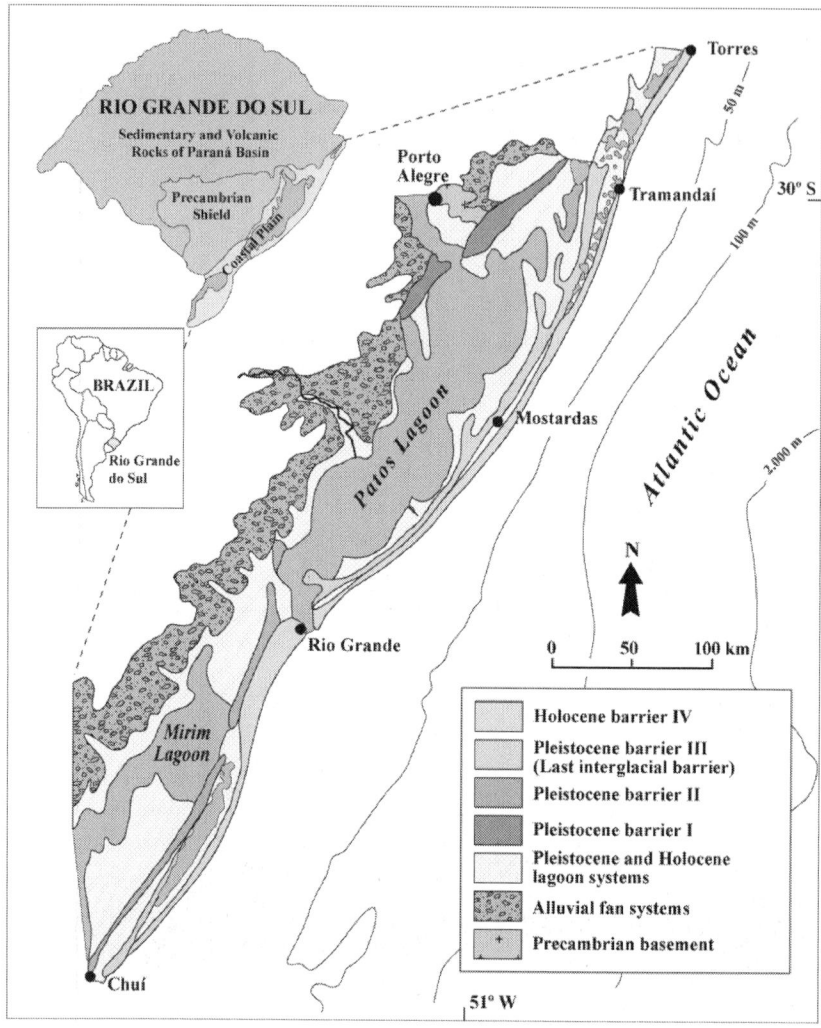

Figure 2. Map of Rio Grande do Sul coastal plain, showing its main geological units (Tomazelli & Villwock, 1996).

The youngest system (IV) is a record of the last glacial cycle, which evolved after the last glacial maximum about 18 ka ago. This is the best kwon geological unit of the Rio Grande do Sul coastal plain, where most of the coastal population lives. Sea level was placed around 120-130 m below the current level about 18 ka ago (Corrêa, 1995). Thereafter, sea level started to rise, exceeding the current level at 7.7-6.9 ka, and reaching its maximum at about 6 ka (Martin et al., 1979; Angulo & Lessa, 1997). By this time, it reached a high 2 to 4 m above the present sea level

(Dillenburg et al., 2000). Since then, it has begun lowering to the current position (Angulo & Lessa, 1997).

Sea level changes controlled the evolution of depositional systems found in the Rio Grande do Sul coastal plain. The deposits associated to this cycle reflect such variations. The understanding of elements composing coastal environments and how they evolved over the last thousands years may be applied to several purposes.

Dillenburg *et al.* (2000) have studied the Holocenic coastal barrier, defining sectors with divergent behaviors in a scale of centuries to millennia. Some sectors are under erosion, others are stable or quasi-stable, and we also find sectors where the beach is growing. Toldo Jr. *et al.* (2005) showed that 442 km of 621 km of coastline have been eroded. One of the main indicators of this process is the mud outcrop of lagoonal origin in the present beach (figure 3). [14]C dating procedures revealed the age of these muds: 5760±120 ka according to Travessas (2003), 4,330±60 ka according to Tomazelli *et al.* (1998), and 3.5 ka according to Dillenburg *et al.* (2004), indicating that the coastal barrier is regressing in these sectors.

Figure 3. Pictures showing beach erosion indicators. The left picture shows mud outcropping at the beach and the right one the buildings destruction due to sea level rises during storms.

GIS-based analysis of outcrops along with satellite images allowed to clarify some specific characteristics of these sectors. Given their particular locations along the coast, they are frequently associated to the occurrence of great dune fields (figure 4). Dune fields are generally found in the far northeastern of the sectors under erosion, what is related to longshore drift (Tomazelli & Villwock, 1992; Toldo Jr. *et al.*, 2004).

The human occupation of some sectors under erosion already has negative consequences. When an elevation of sea level occurs, as a result of meteorological tides, we face the destruction of residences and infrastruture (figure 3). Sectors under erosion should be treated differently from those under accretion in the planning process. It is absolutely necessary to concern about safety measures, such as the maintenance of buffer zones free of development near the beach.

After the expansion of the coastal zone occupation, energy resources are increasingly required. An alternative adopted along Brazilian southern coast is the installation of wind generators. For determining the better place to install them, we must consider several criteria, including legal and socio-environmental aspects.

Figure 4. Image of Rio Grande do Sul coastal plain (Landsat 7, ETM+, Band 2 - 130° inclination) showing its embayments and projections. Alongshore transportation is indicated. Extensive transgressive dune fields occur mainly in the NE portion of erosion sectors (Rosa, 2010).

According to the Brazilian environmental law, dunes are preservation areas. However, the understanding of dune fields dynamic allows to predict dune behavior. Many dune fields in southern Brazil have been monitored by historical aerial photography. Many of them are endangered. Cidreira dune fields are an example of that (figure 5). This field used to be maintained by a sand supply transported by NE winds. Urban development has blocked sand transportation in this area. Thus, dunes and sand sheets are about to disappear, once there is no alternative sand supply to keep sand dunes in place and the current dunes tend to migrate toward coastal lagoons, away from the beaches. The environmental agency in charge authorized a wind farm installation in this area, once sustainable dune fields no longer exist.

In a general sense, when we apply coastal dynamic knowledge to land use planning, environmental impacts of development are likely to be reduced, and a better balance between human needs and the coastal systems carrying capacity is reached.

Modeling of Urban Sprawl Based on GIS: The Rio Grande City Case, South Coastal Plain

The human activities planning process imply combining multiple criteria, including law, ecological function of environments, and human needs. In this sense, a GIS-based urban sprawl model was built by Farina (2003), using a multi-criteria evaluation as a tool for decision making. Rio Grande was chosen to develop a case study of urban sprawl. It has been urbanized and industrialized in an unplanned manner, resulting in environmental degradation of many valuable ecosystems. Rio Grande is located in the south littoral of the Rio Grande do Sul coastal plain (Figure 1). The urban area is physically limited by water, once it is surrounded by the Atlantic Ocean on the east and by the Patos Lagoon on the north and west. Given the physiographic characteristics of this municipality, it has a great potential to port and industrial development, which in turn drives urban expansion.

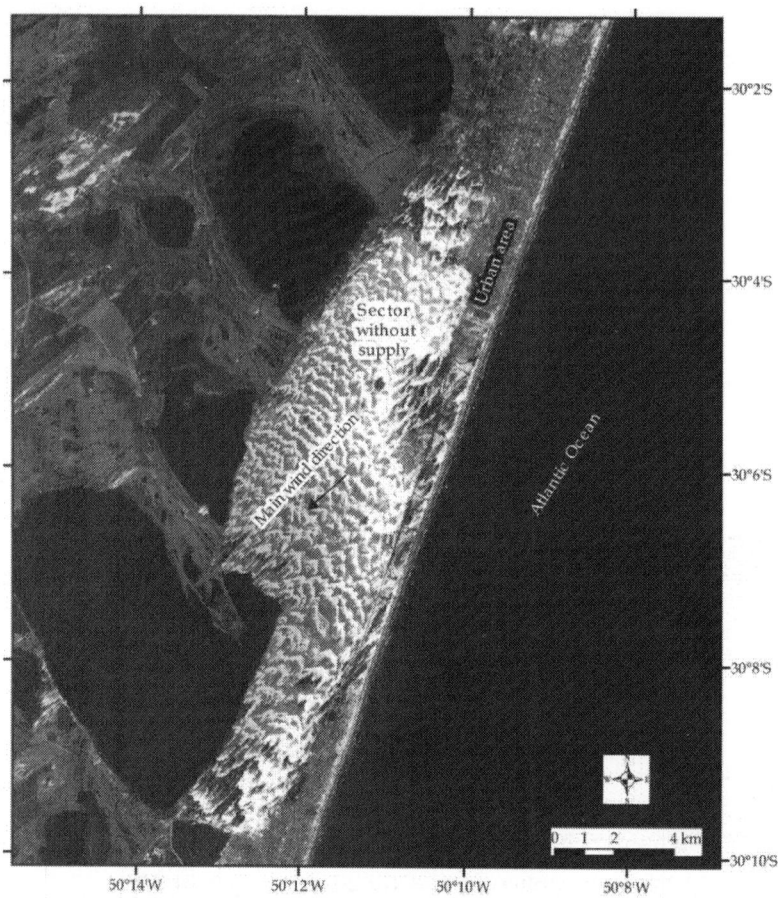

Figure 5. Image of Cidreira dune field (Landsat 7, ETM+, R3G2B1 composition, fused with panchromatic band) showing the northern area where sand supply was canceled by urban occupation.

Decision theory is concerned with the logic by which one arrives at a choice between alternatives. A criterion is some basis for a decision that can be measured and evaluated. Criteria can be of two kinds: factors and constraints, and can pertain either to attributes of the individual or to the entire data set (Eastman *et al*, 1995). A criterion can enhance or detract the suitability of a specific alternative for a given activity. Constraints limits the alternatives under consideration, usually expressed in the form of Boolean maps. Factors, on the other hand, are commonly measured in a continuous scale of suitability. Decision rule is the procedure by which criteria are selected and combined to arrive at a particular evaluation (Gómez & Barredo, 2005, *apud* Eastman *et al*, 1993).

For the Rio Grande urban development model, the following criteria were adopted: proximity to existing urban areas; proximity to the road network; environmental and cultural function of vegetation types; occurrence of flood areas; and occurrence of areas legally constrained to urban development. The geographic data used are presented in the Table 1. Details about the satellite images classification procedures can be found in Farina (2003) and Tagliani (1997). Criteria defined based on the proximity to target areas were obtained through by distance images, where each pixel records the distance to the nearest target feature.

Table 1. Variables and layers relevant to urban occupation.

Variable	Layers	Source
Political boundary of the municipality	Study area	Rio Grande master plan
Urban area	Urban area consolidated; distance from the urban network consolidated	RGB543 color composition of Landsat7 ETM+ images
Road network	Main road; distance from the main road	Topographic maps, updated based on RGB543 color composition of Landsat7 ETM+ images
Vegetation	Vegetal cover	Classification of Landsat7 ETM+ images (Maximum Likelihood Method)
Area of historical and ecological interest	Active dunes, steady dunes; lands recently emerged; swamp; archaeological sites	Rio Grande master plan
Hydrography	Lakes, lagoons, canals, streams, coves	Topographic maps updated based on RGB543 color composition of Landsat7 ETM+ images
Legislation	Distance from water resources; vegetation protected by law; master plan; areas with slope above 30%	Topographic maps Rio Grande master plan RGB543 color composition of Landsat7 ETM+ images
Geology Geomorphology	Flooded areas; areas subject to seasonal flooding; stable substrate	Geological/geomorphologic map produced by Tagliani (1997)

The following criteria were defined as constraints: a) occurrence of urban areas already consolidated; b) water bodies and water courses; and c) occurrence of preservation areas. The following criteria were defined as factors: a) proximity to urban areas and road network; b) environmental and cultural function of vegetation types; c) flood risk; and d) geologic/geomorphologic suitability to urban development.

Constraints were defined by Boolean images, where areas unsuitable to urban development are coded with a 0 and those opened for considering are coded with a 1 (figure 6). Factors, on the other hand, were standardized to a common numeric range (from 0 to 255), and then combined by means of a weighted average. This procedure is known as Weighted Linear Combination (WLC) (Eastman, 1995).

Urban suitability was considered inversely related to the distance from water courses/bodies based on a sigmoidal basis. We choose a sigmoidal function once the need for water resources conservation is greater closer to water bodies/courses, given the higher vulnerability of marginal ecosystems. Such ecosystems tend to be more attractive to irregular settlements. People without access to sanitation and water supply use nearby water bodies for consumption and waste disposal. The initial control point (where the suitability index start to grow) correspond to 30m from water bodies/courses, once this buffer consists in a preservation area, excluded from consideration by legal constraints.

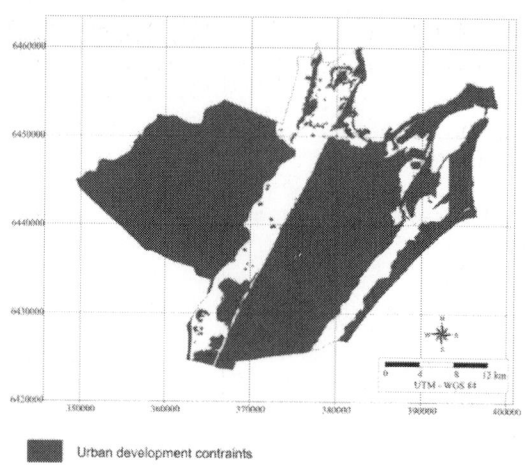

Urban development contraints

Figure 6. Areas constrained to urban development.

The proximity to the urban areas reduces infrastructure installation costs and travel time. Thus, urban suitability is related to distance from urban areas in a linear function. The same was applied regarding the proximity to road network. Control points were set at 0 and 10 km (figure 7).

The weight assigned to factors ranged from 1 to 5.The final result is an image of suitability to urban development, where a value of 0 is assigned to constrained areas and a value ranging from 0 to 255 (highest suitability) is assigned to the areas under consideration, based on a weighted linear combination of factors. To better visualize the results, the urban suitability image was classified as follows: Class I - Very high suitability (170 –255); Classe II - High suitability (170 – 90); Class III – Medium suitability (90 – 40); and Class IV - Low suitability (40 – 0). The urban suitability map is shown in the Figure 8.

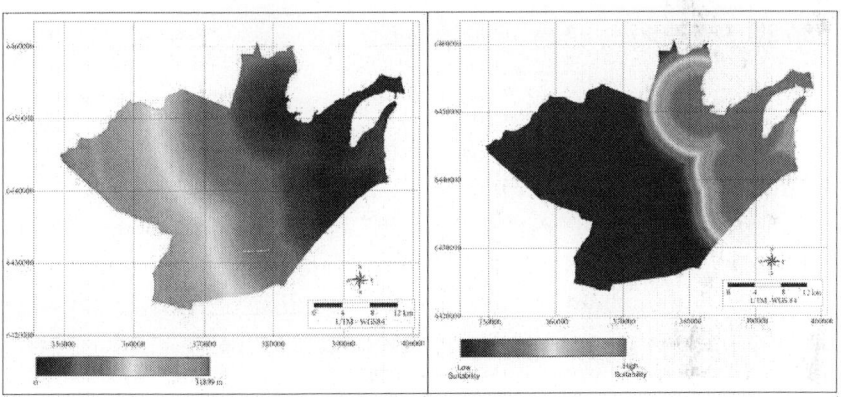

Figure 7. Distance from the urban area and suitability to urban area.

Figure 8. Suitability to urban development.

About 723 km² of the study area, which comprises 926 km²,consists in
constrained areas to urban development. However, unconstrained areas
according to the model are 5.7 times the size of the implemented urban
settlements. The Classes I and II together comprise about 110 km².Based
on the model, they should be priority areas for urban development. Classes
III and IV present lower suitability indexes due to the proximity to water
resources. We recommend a specific occupation plan for these areas,
including, for example but not limited to, public parks and reserves. Total
constraints to urban development represent areas of a high ecological
value. Some of them are legally protected, such as, dunes, native woods,
wetlands, and aquatic ecosystems.

Spatial modeling consists in a important tool to urban development
planning. However, the decision frame adopted for the Rio Grande case
must be adapted to be applied in other coastal municipalities, once the
relative importance of criteria can change from a place to another, as well
as the environmental law at local level.

Land Use Change Detection and Prediction in the Medium Littoral of the Rio Grande Do Sul Coastal Plain

Land use and cover changes are among the major environmental concerns of humanity today, directly or indirectly contributing to climate change, biodiversity loss, and air, soil and water pollution. They are central in the climate change scenario, once they release greenhouse gases, such as carbon compounds (due to deforestation and soil disturbance), methane (due to hydrological alterations, wetland drainage, and rice crop), and nitrogenates (due to the use of fertilizers, irrigation and burning). Besides, the natural vegetation removal by agriculture not only leaves the soil susceptible to erosion, but releases a large amount of nutrients and sediments to adjacent water bodies, causing many negative impacts on environment.

Silva & Tagliani (2012) mapped the land use changes in the medium littoral of the Rio Grande do Sul coastal plain between 1987 and 2000 (figures 9 and 10). The land use and cover change data were used as an input in a change prediction model. Only the main anthropic changes are used in building the model, that is, changes related to deforestation, urbanization, and afforestation. These three processes synthesizes the major land use and cover changes occurred in the region. The software (GIS) interpret them, respectively, as changes from "forest" to "all" classes, from "all" classes to "urban", and from "all" classes to "silviculture".

The change prediction model results in a susceptibility to change image, varying from zero (low susceptibility) to 1 (high susceptibility).The outputs of the model along with an environmental zoning (figure 11), also proposed by Silva & Tagliani (2012) guide decision making and environmental management toward more sustainable actions. The combination (cross tabulation) between susceptibility to change data and the environmental zoning allows to concentrate efforts in specific target areas. The environmental zoning defines the level of protection of land resources while the change prediction model points priority areas (highly susceptible to change) for each management class. Thus preservation areas highly susceptible to change must be the focus of conservative actions, and become priority areas to preventive management. Development areas highly susceptible to change, on the other hand, should also be the focus of actions but as priority areas to infra-structure installation or regulatory measures.

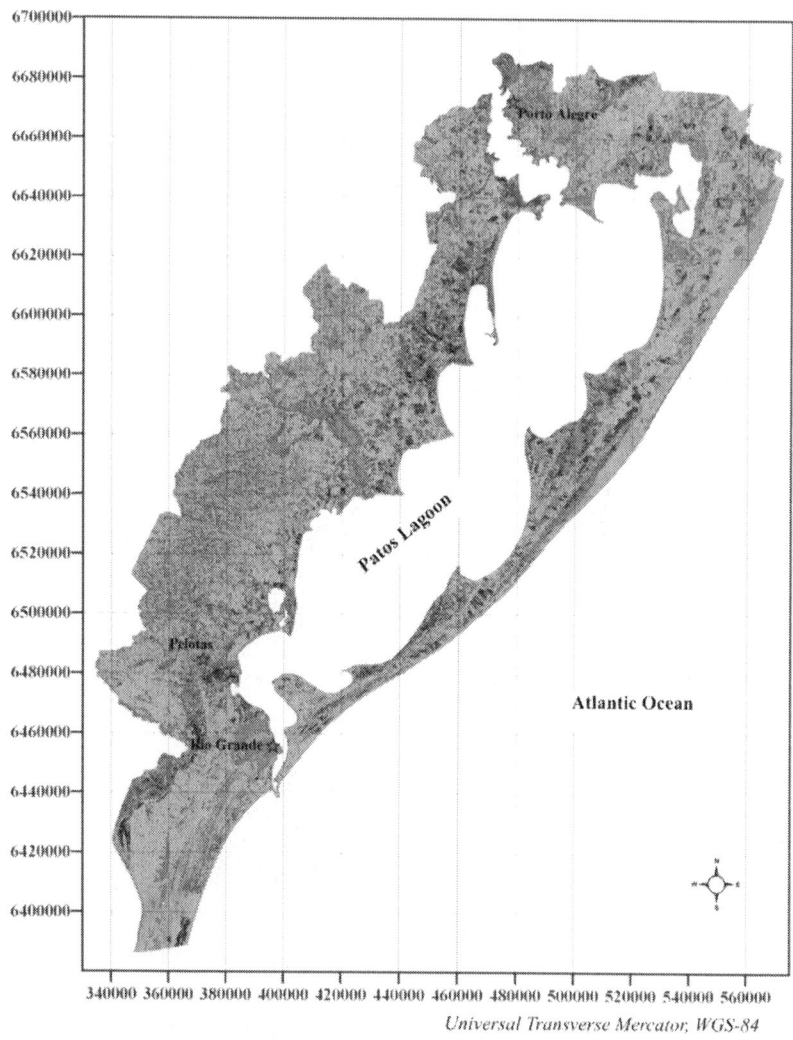

Figure 9. Land use in the medium littoral of the Rio Grande do Sul coastal plain – year 1987.

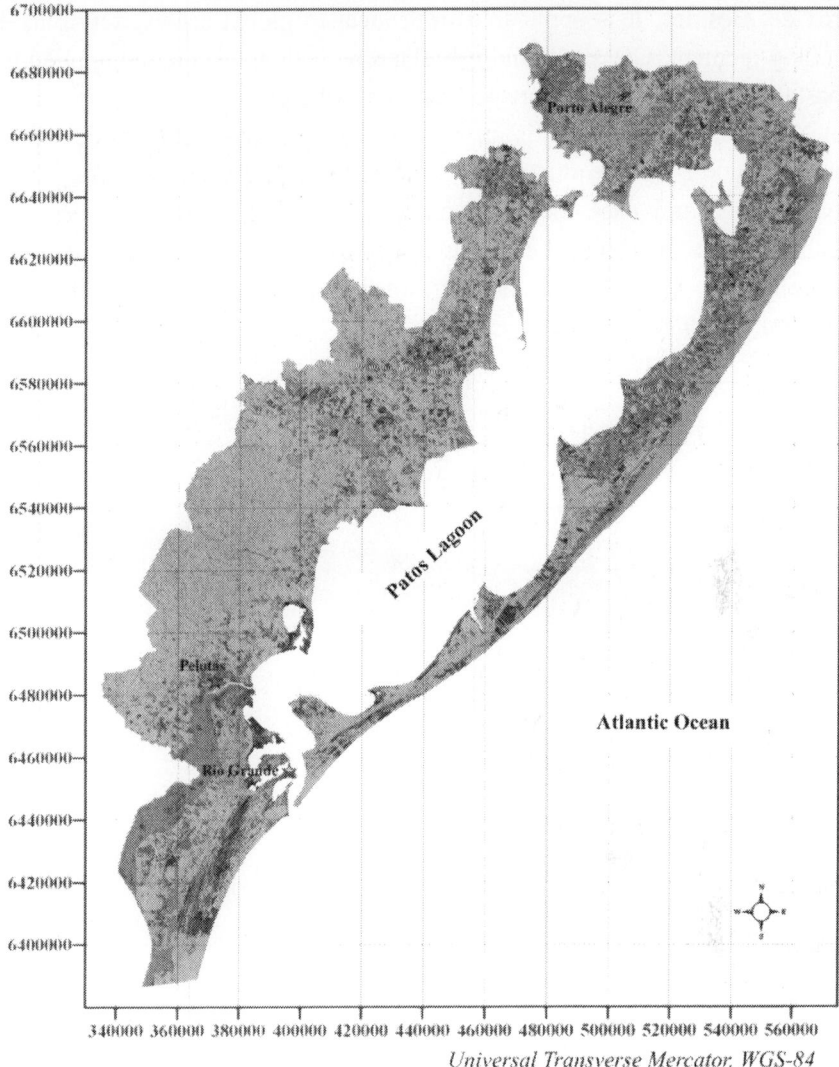

Figure 10. Land use in the medium littoral of the Rio Grande do Sul coastal plain – year 2000.

One of the impediments for the proper development and implementation of the coastal management instruments regards the selection of suitable indicators. Land use (change) seems to be an obvious choice to track several coastal processes, once it can directly or indirectly degrade the health of coastal and marine ecosystems and the services and goods they provide. Besides, it is applicable from local to global scales, acting as a binding element. Classical land use planning techniques can be applied to coastal zone and protected areas, based on static land use and cover maps. The environmental zoning is an example of that. Land change detection and prediction, on the other hand, can properly be used to monitor the coastal zone and base the environmental planning of these areas. In a general sense, land use based studies provide quality information for virtually all coastal management instruments in Brazil (Silva & Tagliani, *op. cit.*).

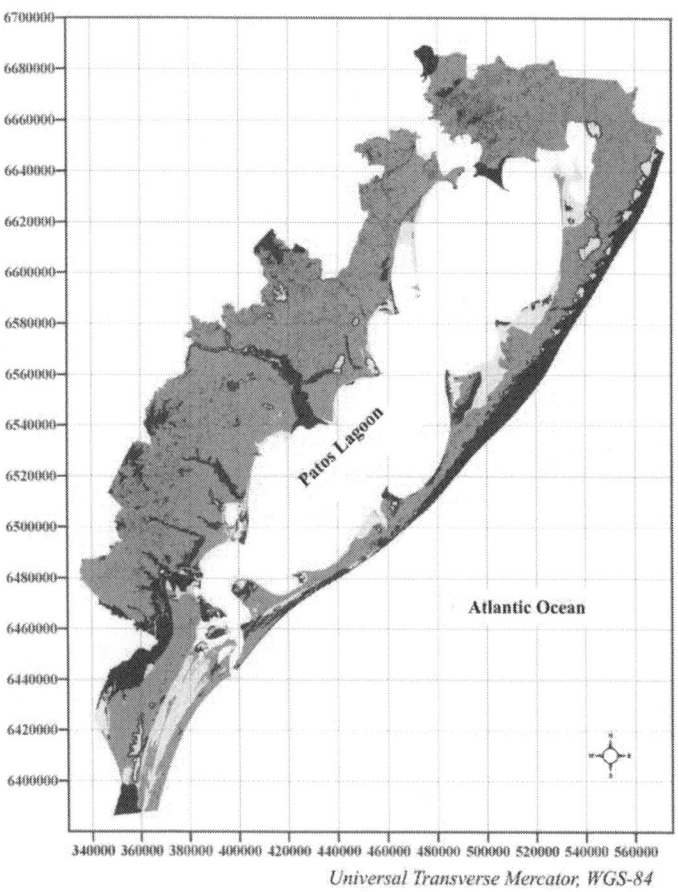

Universal Transverse Mercator, WGS-84

Figure 11. Environmental zoning proposal for the medium littoral of the Rio Grande do Sul coastal plain.

Spatial Modeling Subsides For Risk Management and Coastal Adaptation

The flood and landslide control and prediction are one of our biggest challenges given the current scenario of global changes. Once the medium littoral of the Rio Grande do Sul coastal plain is increasingly urbanized and deforestated, the runoff components are altered and flood and landslide hazards worsen. In this context, GIS has excelled in the development of spatial modeling. Beyond the conventional models, GIS is capable of generating vulnerability and suitability indexes, based on map algebra and context operators, pointing areas at high-risk. Whereas Remote Sensing is an unique data source, GIS spatial modeling allows to establish multiple analytical approaches to assess local vulnerability to environmental changes.

The GIS-based models used to spatially define the relative flood and landslide risk in the medium littoral of the Rio Grande do Sul coastal plain (Silva *et al.*, 2011) incorporates a digital elevation model and also rainfall and soil infiltration potential data to calculate flow direction and accumulation. Other criteria are derived from land use and soil data. Absolute constraints for both models comprise the occurrence of water bodies and wetlands. The resulting images shows the flood and landslide risk, in four classes: low, moderate, high, and very high. About 5% of the area had a decrease in the soil infiltration potential due to land use changes between 1987 and 2000. The flood and landslide risk models gives us a hint about what areas should be the focus of concern in the case of extreme rainfall events. It also can be used as a scenario generation tool, prospecting the effect of land use and climate changes on hydrological vulnerability. Additional criteria, if necessary, can be included to the models. The results are intended to support environmental management and development planning of the costal municipalities surrounding the Patos Lagoon, some of them already suffering the socio-economic consequences of hydro-meteorological disasters. Figure 12 shows the flood risk model results for the years 1987 and 2000.

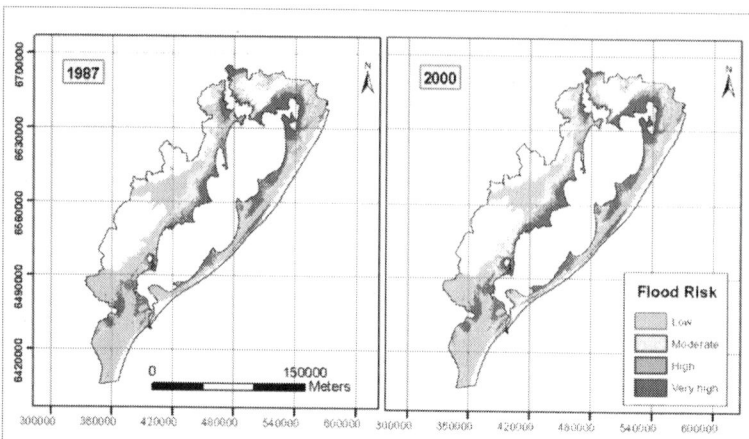

Figure 12. The flood risk images for the years 1987 and 2000.Universal Transverse Mercator, WGS-84.

At least two major hydro-meteorological disasters were recorded in the west coast of Patos Lagoon in the last couple of years. In 2009, 8 people died, 1,200 left homeless, and a bridge was dragged due to an extreme rainfall and flood in Pelotas and Turuçu. In 2011, a similar event resulted in 8 more deaths and 20,000 homeless in São Lourenço do Sul. Half of the urban area was flooded. Two bridges were destroyed and many municipalities isolated. A close view of the model results shows how flood risk responded to land use changes in São Lourenço do Sul, the most impacted city by flood incidents (figure 13).

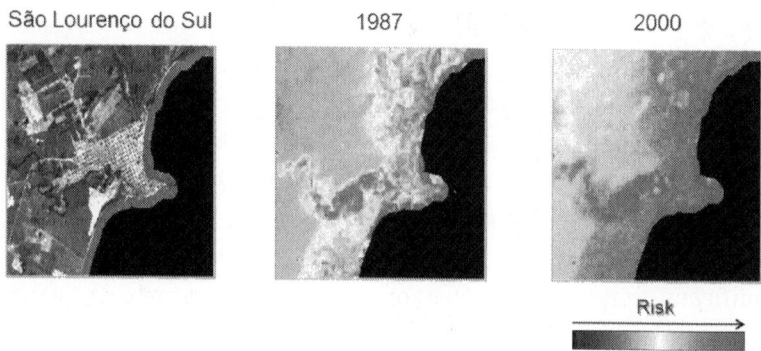

Figure 13. A close view of the flood risk model results for São Lourenço do Sul.

STRATEGIES AND RECOMMENDATIONS TO PROMOTE GIS AS A
KNOWLEDGE TRANSLATION TOOL IN THE REALM OF INTEGRATED
COASTAL ZONE AND OCEAN MANAGEMENT IN SOUTHERN BRAZIL

107

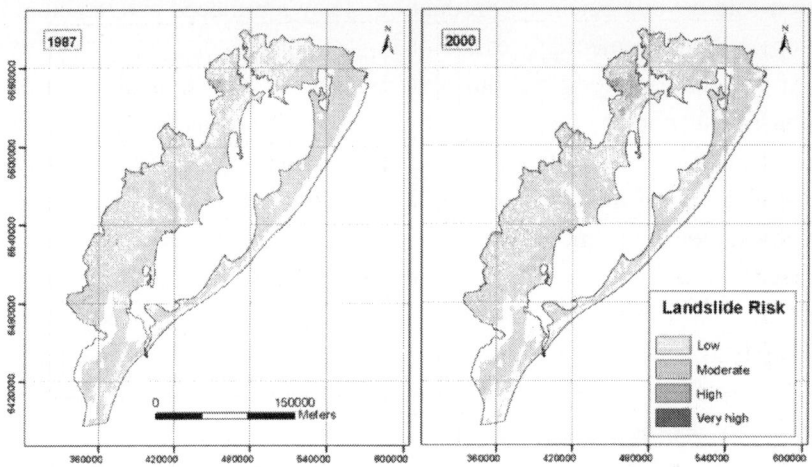

Figure 14. The landslide risk images for the years 1987 and 2000.

Landslides have not affected the region as much as flood, although local incidents have occurred in the last years. However, given the violence of some recent rainfall, combined with land use changes, we may infer about the imminent danger in many parts of the medium littoral. Rio de Janeiro already felt the consequences of this process, which resulted in serious disasters. Thus it is not a surprise if we start facing more severe landslides incidents in the Rio Grande do Sul as well. Figure 14 shows the GIS-based landslide risk results for the years 1987 and 2000.

STRATEGIES AND RECOMMENDATIONS TO PROMOTE GIS AS A KNOWLEDGE TRANSLATION TOOL IN THE REALM OF INTEGRATED COASTAL ZONE AND OCEAN MANAGEMENT IN SOUTHERN BRAZIL

Scientific understanding is crucial for a good ICM decision making (Cicin-Saint *et. al.*, 1998). There is a necessity to establish a relationship between science and management and to incorporate scientific information into the management process. This is especially true in Brazil, where academics play a role beyond scientific production. They are indeed active actors in the coastal management process.

If we take a look at the five phases of the integrated coastal management process, we notice that in most of the cases we only took the first step: issue identification and assessment. It is true regarding integrated coastal management as a hole and also true when we consider the coastal management instruments alone. Even if we consider the I3Geo a successful initiative, we do not have evidences that it has supported the start of new cycles of management, where priorities and policies are adjusted to reflect experience and changing in social/environmental conditions. Besides, the platform works as an one-way tool, giving people access to cartographic information, but it does not promote the communication among sectors of society. The development of web GIS containing public participation tools is a way to overcome this limitation, promoting community empowerment through demand-driven, user-friendly and integrated applications of geoinformation.

The environmental management in Brazil is highly based on licensing. Thus, environmental licensing forces the information exchange between scientists, developers, and environmental agencies. On one hand, licensing is positive, driving research toward real social demands and promoting information interchange. On the other hand, change prediction and scenario generation studies are not required in the licensing process and, consequently, preventive planning is not encouraged. The inclusion of prospective studies in the policy instruments would enhance the efficiency of coastal management, allowing to focus actions on priority areas for development or conservation.

CONCLUSION

Academy is not responsible for defining the planning process, but it plays a major role in supporting it in Brazil, producing relevant information for the decision making process. Research results of some case studies were presented in this sense. Land changes, vital areas suppression, rural impoverishment, urban swelling, they are nothing else but governance failures.

Once the Brazilian National Plan of Coastal Management instruments are spatial in nature, the GIS-based research works have already been helpful in improving the planning process in the Rio Grande do Sul coastal plain. Besides, the acquired knowledge in spatial modeling can be included into the existing instruments to promote their adaptation to the current pace and reach of human activities over the coastal zone.

Geological and geomorphologic mapping of coastal plain is presented here as the basis for land use planning. When coastal dynamic knowledge is applied to land use planning, environmental impacts of development are expected to be reduced.

Geotechnology is also important to urban development planning. The GIS-based model of urban sprawl of the Rio Grande city is an example of that, giving subsidies for management at local level.

Change detection and predictive modeling arise as a vital mean to support the adaption to the current environmental scenario of fast changes. Land use change and environmental risk models seem to bring significant advances in this sense.

A long journey must be undertaken before GIS products and recommendations produced by universities become actions. A new range of opportunities and challenges opens up, whether in the GIS-based environmental plans development and implementation or in further research in spatial modeling as a subside for risk management and coastal adaptation.

REFERENCES

1. R. J. Angulo, G. C. Lessa, 1997GIS Applied to Integrated Coastal Zone and Ocean Management: Mapping, Change Detection and Spatial Modeling for Coastal Management in Southern BrazilMarine Geology140141166

2. Asmus, H.E., Garreta-Harkot, P.F., Tagliani, P.R.A. 1988. Geologia ambiental da região estuarina da Lagoa dos Patos, Brasil. Proceedings of VII Congresso Latino-Americano de Geologia. Belém. November 1988

3. B. Cicin-Sain, R. W. Knecht, G. Kullenberg, 1998GIS Applied to Integrated Coastal Zone and Ocean Management: Mapping, Change Detection and Spatial Modeling for Coastal Management in Southern BrazilIsland Press. 1-55963-604-1

4. I. C. S. Corrêa, 1995Les variations du niveau de la mer durant les derniers 17.500 ans BP: l'exemple de la plate-forme continentale du RioGrande do Sul-Brésil. Marine Geology, 130163178

5. S. R. Dillenburg, P. S. Roy, P. J. Cowell, L. J. Tomazelli, 2000GIS Applied to Integrated Coastal Zone and Ocean Management: Mapping, Change Detection and Spatial Modeling for Coastal Management in Southern BrazilJournal Coastal Research, 167181

6. S. R. Dillenburg, L. J. Tomazelli, E. G. Barboza, 2004GIS Applied to Integrated Coastal Zone and Ocean Management: Mapping, Change Detection and Spatial Modeling for Coastal Management in Southern BrazilMarine GeologyAmsterdan, 2034356

7. J. R. Eastman, 1995GIS Applied to Integrated Coastal Zone and Ocean Management: Mapping, Change Detection and Spatial Modeling for Coastal Management in Southern BrazilClark University, Worcester, USA.

8. J. R. Eastman, W. Jin, P. A. K. Kyem, J. Toledano, 1995Raster procedures for multicriteria/multi-objective decisions. Photogrammetric Engineering and Remote Sensing, 615May 1995), 5395470-09911-129-5

9. F. C. Farina, 2003Utilização de técnicas de geoprocessamento para seleção de áreas adequadas à expansão urbana: caso do município de Rio Grande-RS. UFRGS, Porto Alegre, Brasil.

10. M. D. Gómez, J. I. C. Barredo, 2005GIS Applied to Integrated Coastal Zone and Ocean Management: Mapping, Change Detection and Spatial Modeling for Coastal Management in Southern BrazilMA, 8-47897-673-6Spain.

11. S. Jablonski, M. Filet, 2008Coastal management in Brazil- A political riddle. Ocean & Coastal Management. 51, 5365430964-5691

12. L. Martin, K. Suguio, J. M. Flexor, 1979Le Quaternaire marin du littoral brésilien entre Cananéia (SP) et Barra de Guaratiba (RJ). Proceedings of International symposium of coastal evolution in the Quaternary, São Paulo, Brasil, 296331

13. M. L. C. C. Rosa, 2010Estratigrafia de Sequências: aplicação das ferramentas na alta frequência. Um ensaio na Planície Costeira do Rio Grande do Sul. Ph.D Qualifiyng. Instituto de Geociências. Universidade Federal do Rio Grande do Sul. Porto Alegre, Brasil. 67 p.

14. T. S. Silva, D. De Freitas, P. R. A. Tagliani, F. C. Farina, R. N. Ayup-Zouain, Land use change impact on coastal vulnerability: subsidies for risk management and coastal adaptation. Proceedings of CoastGIS 2011. Ostend. September 2011

15. T. S. Silva, P. R. T. Tagliani, 2012Environmental planning in the medium littoral of the Rio Grande do Sul coastal plain- Southern Brazil: elements for coastal management. Ocean & Coastal Management, 5920300964-5691

16. C. R. A. Tagliani, 1997Proposta para o Manejo Integrado da Exploração de Areia no Município Costeiro de Rio Grande- RS. Um Enfoque Sistêmico. UNISINOS, São Leopoldo, Brasil.

17. P. R. A. Tagliani, H. Landazuri, E. G. Reis, C. R. Tagliani, M. L. Asmus, A. Sánchez-Arcilla, 2003GIS Applied to Integrated Coastal Zone and Ocean Management: Mapping, Change Detection and Spatial Modeling for Coastal Management in Southern Brazilean & Coastal Management. 46, 8078220964-5691

18. Toldo Jr, E. E. , L. E. S. B. Almeida, J. L. Nicolodi, L. R. S. Martins, 2005Retração e Progradação da Zona Costeira do Estado do Rio Grande do Sul. Gravel, 33138

19. L. J. Tomazelli, J. A. Villwock, 1989Processos erosivos na costa do Rio Grande do Sul, Brasil: evidências de uma provável tendência contemporânea de elevação do nível relativo do mar. Proceedings of Congresso da Associação Brasileira de Estudos do Quaternário 2th, 16Rio de Janeiro, Brasil.

20. L. J. Tomazelli, J. A. Villwock, 1992Considerações Sobre o Ambiente Praial e a Deriva Litorânea de Sedimentos ao Longo do Litoral Norte do Rio Grande do Sul, Brasil. Revista Pesquisas, Porto Alegre, 19312

21. L. J. Tomazelli, J. A. Villwock, 1996Quaternary Geological Evolution of Rio Grande do Sul Coastal Plain, Southern Brazil. Anais da Academia Brasileira de Ciências, 683373382

22. F. A. Travessas, 2003Estratigrafia e evolução no Holoceno Superior da barreira costeira entre Tramandaí e Cidreira (RS).Instituto de Geociências, Universidade Federal do Rio Grande do Sul. Porto Alegre, Brasil.

CITATION

Tatiana S. da Silva, Maria Luiza Rosa and Flávia Farina (2012). GIS Applied to Integrated Coastal Zone and Ocean Management: Mapping, Change Detection and Spatial Modeling for Coastal Management in Southern Brazil, Application of Geographic Information Systems, Dr. Bhuiyan Monwar Alam (Ed.), ISBN: 978-953-51-0824-5, InTech, DOI: 10.5772/47840.

CHAPTER 5

Remote Sensing of the Ecology and Functioning of the Mekong River Basin with Special Reference to the Tonle Sap

Simon Nicholas Benger

Flinders University, Australia

INTRODUCTION

The management of large transnational river basins is subject to a range of challenges stemming from differing national priorities, governance of land use activities and resource use, and differences in institutional capacity, data gathering and data sharing. Over vast, often inaccessible areas, remote sensing allows for rapid assessment of ecological resources and hydrological processes. This includes quantification of the extent and ecological functioning of vegetation communities, defining the distribution, duration and timing of flooding, measurement of water quality parameters, groundwater assessment, habitat assessment, and predictive modelling of the ecological impacts of landuse activities and changes to hydrological cycles. Remote Sensing technologies currently allow unparalleled capability for environmental monitoring and management. Data recording and delivery systems, sensor platforms, and sensor technology are constantly improving and each year deliver better remote sensing products for a wide array of applications. Largely independent of geopolitical constraints and boundaries, remote sensing systems allow investigation and analysis of water resources and ecosystem functioning and processes at a range of scales. Large transnational river basins such as the Mekong River basin, can be studied in their entirety or in part.

This chapter examines the use of remote sensing techniques in various investigations in the Mekong River Basin, with particular reference to work on the Tonle Sap (Great Lake) of Cambodia.

The Mekong River Basin

The Mekong is the 10th largest river basin in the world in terms of mean annual outflow, with an annual discharge of 475 billion m^3 (Daming, 1997). From its source on the Tibetan Plateau, it flows some 4,800 km south to the Mekong Delta in Vietnam, draining a total catchment area of 795,000 km^2(MRC, 2005). The Mekong River Basin spans the six countries of China, Myanmar, Lao PDR, Thailand, Cambodia and Vietnam and forms the major hydrological resource for Southeast Asian. The basin has always faced the challenges of widespread poverty, increasing demands on water and environmental resources, and conflict throughout the region (Jacobs, 2002). There is lack of coordinated management of the basin, although the Mekong River Commission (MRC), and its predecessors the Mekong Committee and the Mekong Interim Committee have sought to foster dialogue between the member countries since the late 1950s. The main achievement of the MRC, however, has been the development in recent decades of an extensive data gathering and dissemination system, flood forecasting and warning systems, and advancing the understanding of the ecological and physical attributes of the basin (Jacobs, 2002).

Flow and runoff in the Mekong is strongly seasonal, reflecting the influence of the annual monsoon in the lower reaches of the basin. The wet season peaks in September-October with flows in the lower basin of 20,000-30,000 m^3s^{-1}, compared to dry season flows of approximately 2,000 m^3s^{-1}, which are derived mainly from snow melt in the upper basin (Mekong Secretariat, 1989). The Mekong is subject to natural annual variability which affects the size of the flood peak in any given year and is driven primarily by El Nino Southern Oscillation (ENSO) events (Kiem et al. 2004). Future flood pulse activity may be threatened, however, with significant water resources development occurring throughout the Mekong basin, along with the uncertain effects of climate change on precipitation and river flows. Development and water impoundment and extraction upstream on the Mekong, particularly in southern China but also in Laos,

Thailand and Vietnam, is thought to be affecting the size, timing and intensity of the monsoonal flood pulse (Blake, 2001; Osbourne, 2006). Although catchments in China account for approximately one fifth of the flows in the Mekong overall, they can contribute 70-80% of flows during the dry season (MRC, 2005). The two main dams built by China on the upper reaches of the Mekong are the Manwan dam, which was completed in 1993, and the larger Dachaosan dam, which was completed in 2003. Campbell et al. (2006) show a reduction in average flood height and flooded area over the past decade. One of the most significant hydrological features of the lower reaches of the Mekong basin is the Tonle Sap lake in Cambodia, which fills annually and plays an important role in flood attenuation and sediment and nutrient exchange from the Mekong (MRC, 2005). Events occurring in the upper reaches of the Mekong that systematically alter the flood hydrograph or change its timing are likely to have significant effects on the sustainability of the Tonle Sap (Kummu et al. 2004).

The Tonle Sap

The Tonle Sap or Great Lake of Cambodia (Figure 1) forms part of a unique and ecologically significant sub-system within the Mekong basin. It is the largest freshwater lake in Southeast Asia, covering an area of 250,000-300,000 Ha during the dry season and up to 1.6 million Ha during the wet season (ADB, 2002). Expansion of the lake during the wet season is due primarily to the annual monsoonal flood pulse moving down the Mekong and entering the lake through the Tonle Sap River, which reverses its course as the water level in the Mekong rises above that of the lake. Besides drainage from the Mekong during the monsoonal flood, 13 other catchments drain into the lake. The lake plays an important role in flood peak attenuation and flow control to the Mekong Delta, storing up to 40 km^3 of Mekong floodwater each year and releasing it slowly back into the system (MRC, 2005). It was listed as a UNESCO Biosphere reserve in 1997, and is designated as a Protected Area under Cambodian Royal decree and through numerous international agreements. By far the largest area of savannah swamp forest and inundated forest in Asia, it contains important Ramsar-listed wetlands, and supports extensive fisheries and agriculture of critical importance to the Cambodian economy. Some 2.9 million people live in the five provinces surrounding the lake (ADB,

2002). With economic and political stability returning to the region in the past decade, the population around the margins of the lake is expanding rapidly, along with agricultural activity. Floodplain hydrology and wetland, flooded forest and riparian communities are being modified at a rapid rate and with major ecological impacts.

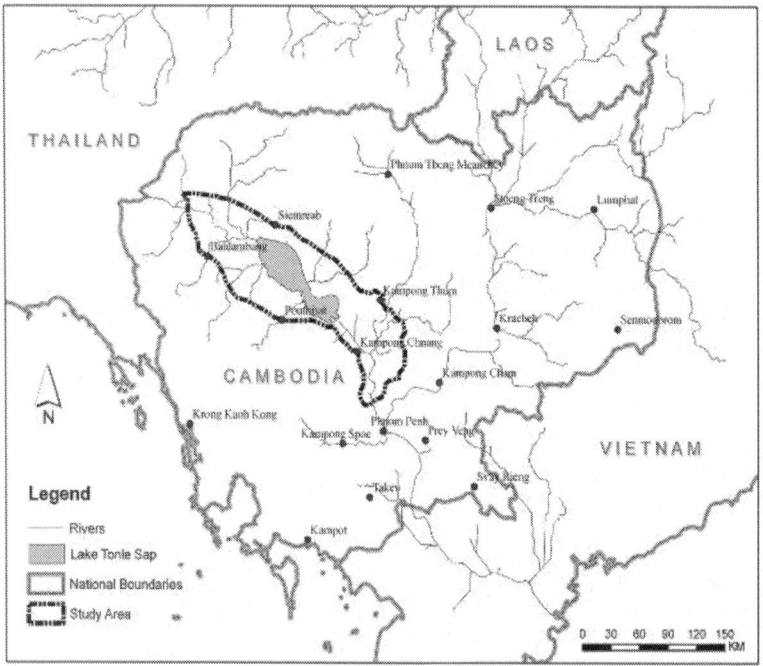

Figure 1. The Tonle Sap Floodplain Study Area.

Several ancient Ankorean capitals flourished on the northern margins of the Tonle Sap floodplain between A.D. 802 and 1431 (Chandler, 1996) and the floodplain has been modified in the past. Structures were built to control the movement of water across the floodplain and harvest it for agricultural purposes or to provide sites for aquaculture activities. This resulted in an extensive hydraulic network estimated to cover more than 1000 km^2 (Evans et al. 2007). Centralised control under the Ankorean court and highly organised agricultural production across the lowlands around Angkor produced economic surpluses for the state. Construction of reservoirs and channels occurred on a large scale through controlled use of labour, including slaves (Higham 2001). Although much of this original irrigation and agricultural infrastructure has probably been subsumed into more recent schemes, or obliterated by the flooding cycles of the lake, several examples of the largest ancient structures remain. These include the Domdek channel: a 200m wide channel extending approximately 80km through the floodplain with 10-15 m high walls; and the Western Baray at Angkor; a water storage covering 17.5 km^2.

The inflow from the Mekong accounts for approximately 70 % of flow into the Tonle Sap lake (Penny, 2006), with the remainder coming from local catchments. Some 80 % of the sediments and nutrients entering the lake from the Mekong are retained (MRC, 2005) and this annual process supports floodplain and fisheries productivity. The Tonle Sap lake is therefore highly susceptible to changes in the size, timing and duration of the annual monsoonal flood pulse, whether that occurs as a result of climate change or upstream water resources development. The past decade has seen reductions in flood height and flooded area of the lake (Campbell et al. 2006), although 2008 saw larger than normal floods throughout the Mekong. Kiem et al. (2008) in their latest modelling, suggest that precipitation will increase by 4.2% on average throughout the Mekong basin, concentrated in the upper sections of the basin in China. Chinvanno (2003) suggests that while there will be some shift in the timing of the flood peak, flooding durations will still be adequate for the survival of significant wetland areas on the Tonle Sap.

Most management efforts on the Tonle Sap to date have focussed on maintaining the lake's fisheries, which provide up to 70% of the protein intake for the entire Cambodian population (van Zalinge et al. 2000), and protection of the Ramsar wetlands as bird nesting sites. Natural resource management is severely under-resourced and occurs in a piecemeal manner (Bonhuer and Lane, 2002) in the face of poorly delineated jurisdictions and conflicting economic interests. Despite the importance of the Tonle Sap lake to the Cambodian economy, only in recent years have authorities and research agencies begun to characterise the flooding cycles of the lake or map floodplain vegetation distributions. Some modelling of lake hydrology was completed in 2003 (Koponen et al. 2003) and an Asian Development Bank project is currently underway to produce GIS datasets of lake resources (ABD, 2002). The Cambodian Mekong National River Commission (MNRC) in association with the multi-country Mekong River Commission (MRC) now monitor flood conditions in the Mekong and the Tonle Sap, but data is restricted to a limited number of gauging stations and is often not reliable. For example, the nearest MRC gauging station is located at Kampong Chhnang, on the Tonle Sap tributary (Figure 1).

REMOTE SENSING OF FLOODPLAIN STRUCTURES

Many extensive water impoundment structures as part of irrigation schemes have been built throughout the Tonle Sap floodplain to retain flood waters and support dry season rice cropping. Such anthropogenic modification of the floodplain occurs primarily on the northern margins of the lake in closer proximity to larger settlements. It is likely that these structures have a significant impact on floodwater distribution and movement and will simply serve as flood barriers if peak lake levels are diminished. Floodplain structures result in permanent inundation of large areas that were previously subject to a wetting and drying cycle; essential for the maintenance and survival of many plant and animal species, including many economically important fish species. In addition, retention and restriction of floodwater movement inhibits nutrient exchange between the floodplain and the lake, and movement of juvenile fish into the lake and the Mekong. The impoundments disrupt the moving littoral of the lake's flood pulse (Junk et al. 1989) where high turnover rates of organic matter and nutrients occur. The gradient of plant species adapted to seasonal degrees of inundation, nutrients and light no longer experiences the conditions under which it evolved.

An aim of the current study was to use remote sensing to determine the extent of floodplain structures around the Tonle Sap and where they lay in relation to flooding extent and duration. Major structures associated with irrigation schemes located within the annually flooded zone of the floodplain were mapped using WAAS corrected GPS to an accuracy of 2-3 m during fieldwork in 2006. High resolution Japanese/NASA ASTER (Advanced Spaceborne Thermal Emission and Reflection Radiometer) imagery was obtained over the floodplain for a range of wet and dry-season dates. ASTER senses in 14 spectral bands in the visible, shortwave and thermal infrared, at 15 m, 30 m and 90 m resolutions respectively (Lillesand et al. 2008). From the 37 ASTER multi-spectral surface reflectance product images obtained for the study, a mosaic of 11 dry-season images covering the Tonle Sap floodplain was constructed with rectification carried out using GCPs (Ground Control Points) collected during fieldwork. The available ASTER coverage over the Tonle Sap is fragmented, both spatially and temporally, due to almost perpetual high levels of cloud cover, but it was possible to generate a near-complete mosaic (Figure 2). As most of the structures occurred on the northern shore of the lake, generally they tended to have an east-west orientation. Horizontal spatial filtering was carried out on the imagery to identify and map the extent of major structures. Spatial filters operate on an image to emphasise or deemphasize image data of varying spatial frequencies. Directional first differencing is a simple directional image enhancement technique which improves the delineation of linear features (Lillesand et al. 2008).

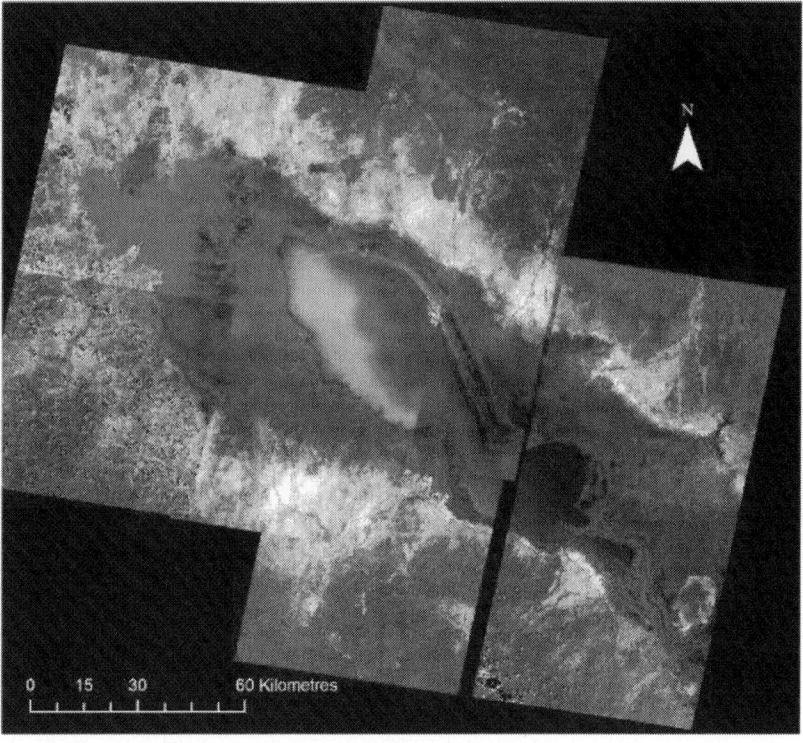

Figure 2. ASTER Dry Season Image Mosaic of the Tonle Sap Floodplain.

Using the filtered images it was possible to identify and map approximately 321 km of major impoundment structures which directly affect water movement across the floodplain (Figure 2). These were generally constructed parallel to the lake shoreline and serve to retain large volumes of water behind them as the lake waters recede after October in any given year. Major structures are defined as being greater than approximately 2m in height, although there are large networks of smaller formal and informal dykes, weirs and regulators which are also used or have been used to modify water movement. Most were built by hand during the Khmer Rouge years using forced labour, and in the absence of any hydrological or engineering knowledge (Kiernan, 1996). Extensive colonisation of these structures with floodplain vegetation has meant that they now form permanent features on the floodplain. According to the flood cycle patterns revealed by the time series analysis described later in this chapter, most of the impoundment structures are built within the zone that would normally be inundated around the end of August in any given year, drying out by mid-December, giving a flood residence time of around 3-4 months (Figure 3). There is also an obvious interaction with floodplain soils. Significant waterlogging occurs around these structures for much of the year, which is a commonly observed phenomenon associated with water storages (Ramireddygari et al. 2000). This is causing a number of changes to wetlands in these areas. Euphorbiaceae, Fabaceae, and Combretaceae species, which once colonised the mosaic of flooded savannah forest are being replaced by those which can tolerate saturated soils. In the areas behind the dyke walls, which now form permanent water storages, natural wetland species have disappeared completely, due either to blanket infestations of water hyacinth and fringing introduced scrub species.

A secondary impact can also be observed. Irrigated rice fields are present on the lake shore side of most water impoundment structures. Increased nutrient levels associated with the application of fertilisers to the rice fields are likely to be affecting the surrounding wetlands through mobilisation during flooding in the wet season and affecting groundwater quality. Leaching of nutrients into the groundwater from these areas, along with increased utilisation of the groundwater by wetland plants due to higher groundwater levels has created succession towards more nutrient tolerant weeds such as *Mimosa pigra* (Campbell et al. 2006). Similarly, pesticides leaching into groundwater which lies close to the surface are affecting the wetland soils which contain the eggs of hundreds of fish species deposited when the lake is in flood. Changes in predator prey relationships that are important for the ecology of the lake (Scheffer, 1998) and its fisheries are likely to be occurring due to floodwater containment. The impoundments would restrict movement of larger fish into shallow areas of lake for predation during flooding and also form barriers to movement of juveniles out from hatchery zones to the lake and the Mekong system. This undoubtedly contributes to the well-documented reduction in the number of fish species and changes in size of individuals (Puy et al. 1999; Bonheur & Lane, 2002). However, the impoundments are also an important source of protein for the occupants of the floodplain, as they effectively operate as large unmanaged aquaculture sites for much of the year, possibly reducing pressure on lake fish stocks.

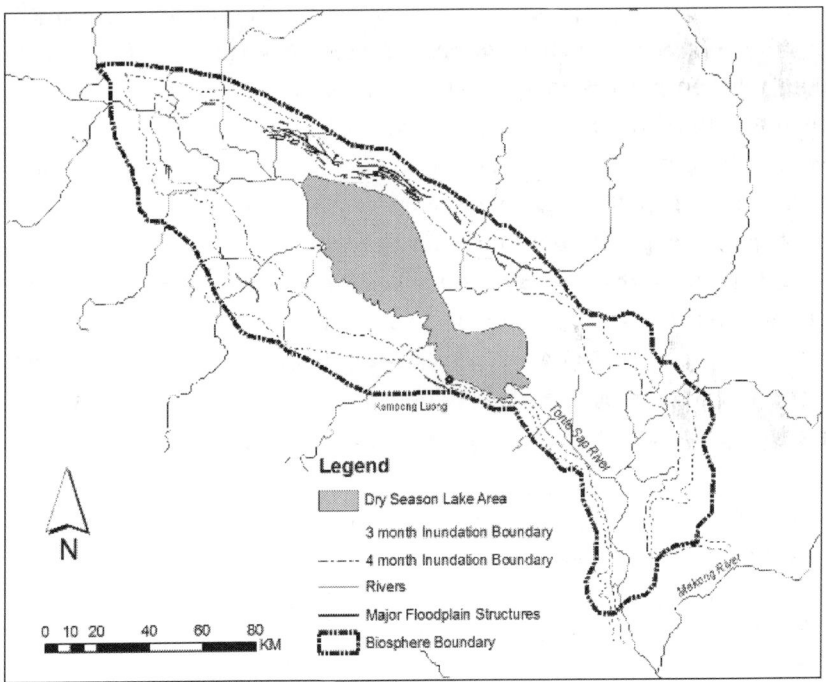

Figure 3. Major Water Impoundment and Barrier Structures on the Tonle Sap Floodplain.

The Khmer Rouge under Pol Pot sought to dramatically increase the areas of land under cultivation on the floodplain, and emptied the cities to provide forced labour for the extensive irrigation schemes that were established (Kiernan, 1996). These structures form by far the largest spatial extent of modifications to the present day floodplain, although many have now been abandoned or are in partial use. Of those surveyed during fieldwork, approximately 40% are now in disuse and others partially used on a seasonally varying basis depending on flooding extent, land availability and population pressures (Bonheur & Lane, 2002). Many of the areas originally modified for rice cultivation have failed to be maintained by the present population because of their inaccessible locations within the floodplain, poor siting, lack of centralised management and maintenance of the schemes, and destruction of infrastructure by flooding. Many of these areas have now reverted to permanent wetlands in areas that would previously have dried out when the floodwaters receded each year. Wright et al. (2004) report on some 570 irrigation schemes existing within the Tonle Sap basin, with only 195 being fully operational today. It is not know how many of these schemes fall within the area of the floodplain, although it is likely that a proportion are located in non-flooded areas. A recent phenomenon on the floodplain is the development of large scale privately owned irrigation schemes which seek to harvest floodwaters for rice production. Substantial areas of floodplain previously utilised by village communes for lower impact agricultural activities are being modified in this way. The use of ring-dyke structures to harvest flood waters for rice production are seen as the ideal new model for agricultural development of the Tonle Sap floodplain (Someth et al. 2009).

REMOTE SENSING OF GROUNDWATER RESOURCES

Remote sensing has been widely used to measure the moisture content of soils (Jensen, 2007), although this often depends on the soil grain size and mineralogy, which will affect the ability of a soil mass to store water. Recently, a number of studies have begun to examine the use of remote sensing for inferring the nature of groundwater resources. Brunner et al. (2007) provide an overview of the potential use of remote sensing in the provision of data to support groundwater modelling in a number of large river basins. Other examples of recent studies include Mutiti et al. (2008), who examined groundwater resource development potential using Landsat imagery, Hendricks Franssen et al. (2008) who inferred groundwater patterning from remotely sensed data, and Milzow et al. (2009) who examined groundwater and hydrology of the large river/wetland system of the Okavango Delta using remote sensing. A range of remote sensing technologies are available to assist in the study of groundwater resources. These include technologies such as radar, LIDAR and digital photogrammetry to derive elevation products, airborne EM (electromagnetics) to examine changes in electrical conductivity in the shallow subsurface, and the remote sensing of vegetation, salt crusts and other surface features as a proxy for subsurface groundwater conditions (Brunner et al. 2007).

Groundwater resources are particularly important for the region in and around the Tonle Sap floodplain, as they form the major water supply for human use (Wright et al. 2004). The sedimentary depression of the Tonle Sap is surrounded by low-lying alluvium, with older coarser ferruginous silts, sands and grits around the perimeter overlain by red-clayey and silty sediments (Stanger et al. 2005). The alluvial deposits of the Tonle Sap floodplain are believed to be very good shallow aquifers, with high recharge rates (5-20 m^3/h) and a groundwater table generally within 4-6m of the surface. Groundwater quality is generally good apart from high iron content reducing palatability in some areas, and dangerous levels of arsenic contamination in others (Wright et al. 2004). In response to the large amplitude floods that characterise the hydrological cycle of the Tonle Sap, there is an annual cycle in groundwater levels from depths of around 6 m in riparian areas to a few centimetres in some parts of the floodplain (Stanger et al. 2005).

Loss of vegetation, particularly deep rooted tree species, reduces uptake of water from the soil profile and exacerbates waterlogging problems in the wetlands. A large seasonal population usually migrates from upland areas and the non-flooded areas of the Tonle Sap basin to the floodplain as the floodwaters recede, building temporary settlements on and around the water impoundment structures (Bonheur & Lane, 2002). The temporary settlements facilitate activities such as dry season rice cropping and fishing and informal aquaculture. Human settlement compounds the loss of larger wetland tree species in these areas as they form the primary source of fuelwood and building materials. This occurs on a wide scale despite a complete ban on all forms of timber extraction from the flooded forest areas. As well as the loss of deep rooted tree species, groundwater levels are also likely to be affected by the permanent and semi-permanent water impoundments, which would have a subsurface connection to the local water table (Ramireddygari et al. 2000). An aim of the current study was to investigate whether these effects existed and were detectable using available optical remotely sensed imagery. Soil moisture absorbs incident radiant energy in the 1.4, 1.9 and 2.7 μm regions, although the spectral response can be complex depending on soil type and soil characteristics (Jensen, 2007).

In three fieldsite locations on the Tonle sap floodplain, the relationship between groundwater and water storages was examined. During fieldwork elevated soil waterlogging adjacent to water storages could be observed through the presence of dark saturated soils along with consequent changes in vegetation type. Trenches were dug adjacent to the structures to ascertain depth to water table, and these confirmed water tables lying at or near the surface. Remotely sensed analysis of Landsat imagery over these areas made it possible to map the extent of waterlogging extending out from these structures. This involved the generation of wetness index maps, using the Kauth-Thomas (KT) transformation (Kauth & Thomas, 1976; Collins & Woodcock, 1996). A wetness index map derived from the KT transformation will indicate not only the level of surface soil moisture, but also the wetness of associated vegetation (Mutiti et al. 2008). A wetness index map for the fieldsite locations examined is presented in Figure 3. While the results do indicate a relationship between the size of the water storage and the area detected, such results are difficult to interpret without further information on the quantity of water stored, the duration of storage, the soil types and localised topography, all of which are unavailable for the Tonle Sap Floodplain. However, they did indicate the potential of remote sensing to detect and quantify these effects, and demonstrate the effects of waterlogging of soils adjacent to water impoundment structures – an important consideration given the rapid agricultural development occurring in some areas of the floodplain utilising water impoundments.

REMOTE SENSING OF FLOODPLAIN VEGETATION

The monsoonal driven flood pulse fills the lake and floods an extensive area of the floodplain, usually for several months from August through to January, creating a unique flooded forest plant community (McDonald et al. 1997). These temporary wetlands serve essential ecosystem processes in terms of nutrient exchange between the lake (and the Mekong system upstream) and the floodplain, and are essential for fish breeding (Puy et al. 1999). Flooded forests are found mainly around the dry-season lake shoreline and comprise about 10% of the floodplain and are dominated by *Barringtonia acutangula, Baringtonia micratha* and *Diospyros cambodiana*. At higher elevations are extensive areas of short tree shrubland dominated by species of Euphorbiaceae, Fabaceae, and Combretaceae, together with *Barringtonia acutangula* (Wikramanayake & Dinerstein 2001) and seasonally flooded sedgelands and grasslands occupy the distal margins. Large seasonal contrasts in lake levels affect the characteristics of the wetland vegetation (Penny, 2006), with some forest areas enduring fluctuations of up to 8m and complete canopy submergence for months at a time (McDonald et al. 1997).

Figure 4. KT Wetness Index Map of an area of the Tonle Sap Floodplain.

Like many ephemeral wetlands around the world, the distribution of the mosaic of flooded forest, scrub and grassland around the lake is determined largely by the duration and depth of flooding (Bonheur & Lane, 2002), and to a lesser degree substrate. The tropical climate, nutrient rich soils and abundant water present on the floodplain mean that vegetative growth occurs rapidly, and forests and wetlands quickly regenerate. It has been suggested that much of the present wetland vegetation is secondary regrowth (McDonald et al. 1997), although this seems unlikely over the majority of the Tonle Sap wetlands, which are highly inaccessible. Only on the northwestern margins of the lake, where the ancient civilizations of Angkor flourished between A.D. 802 and 1431 (Chandler, 1996) is large scale clearing likely to have occurred to facilitate extensive agricultural schemes. Many of these were re-established, often unsuccessfully, during the Khmer Rouge period (Kiernan, 1996). While limited historical data exists on the distribution of plant communities across the floodplain, the majority of the floodplain vegetation is still intact, although often modified in areas closer to settlements, and capable of near normal ecological functioning subject to floodwater availability.

One aim of the current study was to use remotely sensed data to map the current distribution of wetland and floodplain vegetation around the Tonle Sap, so that these could be examined in relation to where they occur spatially on the floodplain in relation to flooding extent and duration. Previous efforts to map wetland vegetation distribution and its relationships to flooding (e.g. Kite 2001) utilised generalised USGS landcover classifications not particularly suited to the Tonle Sap floodplain. The ADB has generated landuse maps and metrics for all of the Tonle Sap catchments (ADB, 2009), although the landuse categories used are generalised and non-species specific.

Image classification procedures can be used to identify, map and quantify vegetation units of interest in remotely sensed imagery. The overall objective of image classification procedures is to automatically categorise all pixels in an image into land cover classes or themes (Lillesand et al. 2008). Image classification attempts to use the spectral information available in the data for each pixel as the numerical basis for categorisation. Different feature types will manifest different combinations of spectral response in each band (depending on sensor type) based on their inherent spectral reflectance and emittance properties which may also be variant in space and time. Spectral pattern recognition refers to the family of classification procedures that utilise this pixel-by-pixel spectral information as the basis for automated land classification (Lillesand et al. 2008).

Supervised classification is the procedure most often used for quantitative analysis of remote sensing image data. It rests upon using suitable algorithms to label the pixels in an image as representing particular ground cover types or classes (Richards & Jia, 2006). The multidimensional normal distribution of a spectral class is specified completely by its mean vector and its covariance matrix. Consequently, if the mean vectors and the covariance matrices are known for each spectral class then it is possible to compute the set of probabilities that describe the relative likelihoods of a pattern at a particular location belonging to each of those classes (Lillesand et al. 2008). It can then be considered as belonging to the class which indicates the highest probability. Therefore, if the mean vectors and the covariance matrix are known for every spectral class in an image, every pixel in the image can be examined and labelled corresponding to the most likely class on the basis of the probabilities computed for the particular location for a pixel. Before that classification can be performed however, the mean vectors and covariance matrix are estimated for each class from a representative set of pixels, called a training set. These are pixels which the analyst knows as coming from a particular spectral class.

Supervised classification consists therefore of three broad steps. First a set of training pixels is selected for each spectral class using the reference data available in the form of digital vegetation maps. The second step is to determine the mean vectors and covariance matrices for each class from the training data. This completes the learning phase. The third step is the classification stage, in which the relative likelihoods for each pixel in the image are computed and the pixel labelled according to the highest likelihood (Richards & Jia, 2006). Numerous mathematical approaches have been developed for spectral pattern recognition and it is beyond the scope and relevance of this chapter to review them all. Some commonly used classifiers are the Minimum-Distance-to-Means, parallelepiped, Gaussian Maximum Likelihood Classifier (MLC) and the Piecewise Linear Classifier. In the current study, MLC was used for the supervised classification of the ASTER optical imagery. MLC has a demonstrated reliability in achieving accurate classification of land cover types across a range of different environments (Bolstad & Lillesand, 1991; San Miguel-Ayanz & Biging, 1997).

MLC is one of the most commonly used supervised classification methods and it has been demonstrated to be extremely powerful and efficient in a great number of investigations (Maselli et al. 1990). It works most effectively when dealing with normal distribution in the spectral data, although it has also been shown to be relatively resistant to class distribution anomalies (Hixson et al. 1980; Yool et al. 1986). This classifier quantitatively evaluates both the variance and the covariance of the category spectral response patterns. It assumes a Gaussian distribution in the category training data, which is generally a reasonable assumption. Using this assumption the distribution of a category response pattern can be completely described by the mean vector and covariance matrix, it is possible to compute the statistical probability of an unknown pixel belonging to particular land cover class. In essence the maximum likelihood classifier delineates ellipsoidal "equiprobability contours" in the scatter diagram of spectral values which act as the decision regions (Lillesand et al. 2008).

The main limitation of maximum likelihood classification is the large number of computations required to classify each pixel. This is particularly true when either a large number of spectral channels are involved or a large number of spectral classes must be differentiated. Numerous extensions and refinements of the maximum likelihood classifier have been developed (Lillesand et al. 2008). These include the use of lookup tables in which the category identity of all possible combinations of digital numbers is determined prior to classifying the image and each unknown pixel is classified simply by reference to these lookup tables. This avoids the need to carry out complex statistical calculations for each pixel as they have already been determined for each category. Another means of optimizing maximum likelihood classifiers is to use some method to reduce the dimensionality of the dataset used to perform the classification. Procedures such as the principal components, canonical components (Jensen and Waltz, 1979) and tassled cap (Kauth and Thomas, 1976) transformations achieve this reduction of the dataset by making use only of the significant sections of the data.

Floodplain vegetation type and distribution were observed and mapped in the field in and around the Tonle Sap through a number of fieldwork surveys conducted in 2005 and 2006, in accessible locations. Remote sensing offers the ability to map landcover types over large areas, based on spectral information collected from representative vegetation communities and other landuse and landcovers (Lillesand et al. 2008). On the basis of training sites mapped throughout the floodplain during fieldwork, wetland vegetation and landcover across the floodplain was classified into 20 classes using maximum likelihood classification on the 9 visible/near-infrared and shortwave infrared bands of the ASTER imagery, which were resampled to 30 m. This facilitated determination of the types and extent of wetland vegetation directly affected by the water impoundment structures and their relationship to flooding patterns. Classification accuracy was assessed using standard confusion matrices to generate overall accuracy and Kappa statistics (Congalton & Green, 2008), using one training site for each landcover type not used in the original classification. As a result of the classification carried out over the floodplain using the imagery, it was possible to generate floodplain metrics for the various vegetation and landuse classes, and these are presented in Table 1.

Table 1. Landcover classification results for the Tonle Sap floodplain.

Vegetation/Landuse Class	Area (ha)	Percentage
Barringtonia acutangula dom.	107928	7.75
Barringtonia acutangula dom.	765737	55.01
Diospyros cambodiana dom.	184344	13.24
Euphorbiaceae Shrubland	53794	3.86
Tiliaceae Shrubland	61062	4.39
Mimosa pigra	7551	0.54
Sedge	17810	1.28
Phragmites Reeds	36928	2.65
Thornbush	3380	0.24
Water storage, unvegetated	19257	1.38
Water storage, vegetated	11345	0.81
Agricultural - rice	34972	2.51
Agricultural - fallow	11839	0.85

Legume cropping	250	0.02
Grasslands	67410	4.84
Mudbanks saturated soil	2722	0.2
Bare dry soil	2344	0.17
Firescar	1569	0.11
Rock outcrop	183	0.01
Human settlement	1664	0.12
Total (excluding)	1392089	100

Kappa = 0.83

RELATIONSHIPS BETWEEN ELEVATION AND FLOODPLAIN VEGETATION

High quality digital elevation data are essential for the assessment of floodplains and spatial arrangement of vegetation communities. Numerous studies of wetland vegetation have suggested that elevation is a primary determinant of vegetation type and location within wetland systems (Scoones, 1981; Hughes, 1990), and substrate to a lesser degree. Analysis of elevation data can yield important information on the spatial arrangement of vegetation in wetland and floodplain environments as it determines the extent and duration of flooding of these areas. In many of the developing countries which comprise the Mekong River basin high quality survey data is simply not available, and over large inaccessible areas such as the Tonle Sap floodplain, ground based survey is logistically impossible. Therefore remote sensing offers the primary means of gathering such data.

There are a range of remote sensing techniques available for the generation of elevation data or digital elevation models (dems). In general, higher precision in these products is accompanied by higher cost of acquisition and processing. Techniques include digital stereo photogrammetry, radar interferometry and LIDAR. For the Mekong basin the primary dataset that has been utilised is the United States Geological Survey (USGS) GTOPO 30 dem, which is a 30 arc-second resolution product. The Shuttle Radar Topography Mission (SRTM) global product can also be used which has a 3 arc-second (approximately 90 m) resolution with 5 m vertical accuracy (Slater et al. 2006), and more recently, the ASTER GDEM global dem became available in 2009 with 30 m resolution and 15 m vertical accuracy. Of these datasets, only the latter is suitable for use in a low relief environment such as the Tonle Sap floodplain. In the GTOPO 30 and SRTM data, variations in floodplain relief are dominated by data anomalies. In all cases where remotely sensed elevation data are available, finer resolution dem data can be interpolated, but these may lead to a false representation of precision as they will normally retain the errors present in the original data (Longley et al. 2007). Kite (2001) used the USGS GTOPO 30 product for hydrological modelling of the Mekong Basin, and the ADB (2009) show flooded area maps and metrics for the Tonle Sap catchments interpolated from contour maps. In the current study, remotely sensed elevation data was utilised to investigate the relationship between the primary wetland and floodplain vegetation types and elevation, and hence relationship to flooding. For this purpose the ASTER GDEM product was used after processing to remove anomalies, most of which occur over areas of open water and along tile edges, and extracting only elevations below 30m in height. The resultant 30m dem for the Tonle Sap floodplain is shown in 3D in Fig 5.

Figure 5. 3D dem of Tonle Sap floodplain, derived from ASTER GDEM.

A simple GIS-based analysis of the location of vegetation communities in relation to elevation yields information on the elevation ranges they occupy within the floodplain. While elevation alone is not the sole determinant of flooding effects on vegetation, in a floodplain such as the Tonle Sap, where overbank flooding from the lake is the primary source of floodwater, it does indicate the sensitivities of various ecological communities to water level ranges. The depth and duration of flooding for these communities is a primary determinant of their evolution in a given location and their ecological functioning (Campbell et al. 2006). The main vegetation classes and their elevation ranges derived from this analysis are shown in Table 2.

Table 2. Elevation ranges for primary vegetation classes.

Vegetation Class	Min imum E levation (m)	Max imum Elevation (m)	(m)
Barringtonia acutangula dom.	0.6	1.8	1.2
Barringtonia acutangula dom.	1.9	8.2	6.3
Diospyros cambodiana dom.	2.4	8.7	6.3
Euphorbiaceae Shrubland	5.3	10.4	5.1
Tiliaceae Shrubland	6.2	12.6	6.4
Sedge	8.2	11.3	3.1
Phragmites Reeds	1	2.6	1.6
Grassland	9.1	18.4	9.3

The elevation ranges of the primary vegetation classes of the Tonle Sap confirm that the flooded forest and reed communities occupy lower elevations on the floodplain, with Savannah woodland communities at higher elevations followed by shrubland, sedge and grasslands. Flooded forest, reed and sedge communities occupy the narrowest elevation ranges on the floodplain, while those communities at higher elevations are most likely to be affected by reductions in flood height. This information can then be used with the information on temporal flood extent patterns described below to characterise the horizontal and vertical arrangement of species on the floodplain. This landscape ecology approach to the understanding of floodplain structure provides important information on ecological functioning. Landscape ecology is based on the hypothesis that the interactions among biotic and abiotic components of the landscape are spatially mediated. Not only are the flows of energy material or species from place to place affected by the locations of the places in the landscape, but these flows then determine the interactions among energy, material and species (Malanson, 1993). A central theme of landscape ecology is that spatial structure controls the processes that continuously reproduce the structure. Landscape ecology is an approach to the study of the environment that emphasizes complex spatial relations. The relative locations of phenomena, their overall arrangement in a mosaic and the types of boundaries between them, become the priorities of study (Forman & Godran, 1986; Ingegnoli, 2002).

FLOOD DETECTION AND MAPPING

The monsoonal flood pulse is the primary mechanism affecting productivity in the Tonle Sap lake, wetlands and floodplain. The economically important fisheries of the Tonle Sap are strongly influenced by the maximum flooded area and resultant area of fish feeding and breeding habitat (Webby et al. 2005). Remote sensing of the inundation patterns across the study area therefore formed an important part of the current study. Knowledge of the extent and residence time of floodwaters on the floodplains of major rivers is essential for hydrological and biological studies of these systems, and yet for most areas of the Mekong, this remains largely unknown beyond simple maps of flood extent. For the Tonle Sap, the ADB has compiled maps showing minimum and maximum flood extents for the catchments around the lake as derived from satellite image interpretation (ADB, 2009). For the areas examined in this study, information on flow rates and stream heights may be available, but because of the low relief and complex hydrology of many wetland areas, these data do not correlate well with inundation patterns. Rates of organic matter production, decomposition and export to the river channel are closely linked to floodplain inundation patterns. Primary production rates in inland wetlands are very high and these communities may cover hundreds of thousands of square kilometres (Matthews and Fung, 1987). In many large river systems with associated extensive wetland areas, the difficulty in determining the extent of flooding makes it difficult to accurately estimate wetland area and characterise vegetation relationships. Ground measurement of flooding in forested wetlands is severely limited by the inaccessibility typical of these areas, where mobility is often hampered by flooding and boggy conditions. Remote sensing offers the ability to detect flooded over such areas, and this is typically done using optical or radar imagery.

With regard to optical remote sensing of inundation and the spectral reflectance of water, probably the most distinctive characteristic is the absorption of energy at near-infrared (NIR) wavelengths. Locating and delineating water bodies with remote sensing data is done most easily at NIR wavelengths because of this property (Lillesand et al. 2008). However, various conditions of water bodies manifest themselves primarily in visible wavelengths. Landsat TM imagery has been used to map floodwater distribution and characteristics (e.g Imhoff et al. 1987; Pope et al. 1992; Mertes et al. 1993, 1995;Johnston and Barson 1993), and optical SPOT data has also been used for floodwater mapping (Blasco et al. 1992).

Remote sensing of flooding may also be hampered by forest canopies that render the land/water boundary invisible to infrared and visible wavelength sensors and by frequent cloud cover during periods of rainfall. These limitations are largely overcome by SAR radar sensors which are unaffected by clouds and can significantly penetrate relatively dense forest canopies (Hess et al. 1990). Passive microwave remote sensing has also proved useful for revealing large-scale inundation patterns, even in the presence of cloud cover and dense vegetation (Choudery 1991, Sippel et al. 1994). The bright appearance of flooded forests on radar images results from double-bounce reflections between smooth water surfaces and tree trunks or branches. Enhanced back scattering at L-band has been shown to occur in a wide variety of forest types and is a function of both stand density and branching structure (Hess et al. 1990). Steep incidence angles (20-30°) are optimal for detection of flooding, since some forests exhibit bright returns only at steeper angles. Backscattering from flooded forests is enhanced by underlying water. For forests of moderate density, L-band returns are dominated by corner reflections between trunks and surface and between branches and surface (Richards et al. 1987a). Scattering from a smooth water surface is specular, whereas that from soil includes a significant diffuse component and therefore the amplitude of returns will be higher for standing water beneath forests.

There is a high degree of structural diversity associated with flooded forests, as they occur on numerous substrates, in both saline and fresh water and at a wide range of latitudes (Matthews and Fung 1987). Most frequently studied have been the swamp forests of the coastal plains of the southeastern United States. Relatively bright L-band returns from semi-permanently to permanently flooded stands have been reported in several studies (e.g. Hoffer et al. 1986, Evans et al. 1986, Wu and Sader 1987). Detection of underlying water in mangrove swamps was demonstrated by Imhoff et al. (1987) in the Sundarbans region of Bangladesh and by Ford and Casey (1988) in East Kalimantan. Bright returns for seasonally inundated temperate forests are described by Richards et al. (1987b) for Eucalyptus camaldulensis forests in Australia. Ford et al. (1986) distinguished flooded varzea forest from non-flooded forest using SIR-B scenes of the Rio Japura in the Amazon Basin.

The forest stands cited above have very diverse structures: canopy depth relative to total tree height, dominant branching angle, and crown shape are quite variable. They also encompass a wide range of leaf type and tree heights. It is clear that stands with low stem densities may appear bright at L-band (Hess et al. 1990). Enhancement has also been shown for stands described as dense or thick (Hoffer et al. 1986, Ford and Casey 1988). Enhanced backscattering from flooded forests thus occurs over a broad range of tree species, canopy structures and stand densities. Richards et al. (1987b) demonstrated that brighter returns from flooded forests are not simply a function of vegetation differences between upland and lowland sites. They were able to clearly distinguish between flooded and non-flooded portions of a single forest type.

The accuracy of flood detection using radar imagery is difficult to determine since most studies of flooded forests focus only on those areas which do yield bright L-band returns. Near or complete absence of backscatter from flooded Maryland swamps with dense canopies has been noted by Krohn et al. (1983). It appears that dense undergrowth may significantly affect double-bounce returns. Ford and Casey (1988) found the opposite to be true, however, in flooded mangrove forests of Kalimantan. They found that open stands of low slender trees did not yield bright returns on SIR-B imagery while adjacent denser mangrove stands did. The above examples suggest that for certain forest types, the extent of flooding beneath the canopy would be underestimated using L-band radar. Overestimation would occur if other targets yielding bright returns were mistaken for flooded forests. Other sources of bright returns would normally be able to be visually distinguished from flooded forest based on shape, pattern, associated features and minimal site knowledge. A more serious source of confusion is non-forest vegetation naturally occurring adjacent to flooded forest. Flooded marshes (emergent herbaceous vegetation) typically appear dark at L-band (Krohn et al. 1983, Ormsby et al. 1985). However, marsh vegetation sometimes yields bright returns very similar to those from flooded forests (Krohn et al. 1983).

The magnitude of enhancement associated with double bounce beneath flooded forests can vary significantly. In many studies, variations in magnitude appear to be the result of differences in stand composition as well as flooding (Hess et al. 1990). The problem of separating backscatter variation caused by differences in vegetation from that caused by flooding was minimised in the study by Richards et al. (1987b), because of the virtually monospecifc stands of eucalyptus examined. Backscattering from flooded and non-flooded sites within the forest was estimated to vary by 10.8 dB: a substantial difference. Treating the canopy as a uniform layer of small particles, Engheta and Elachi (1982) estimate the enhancement resulting from the presence of a perfectly reflecting surface beneath the canopy to be 3 to 6 dB. It appears from the literature that L-band radar imagery used in the current study should enable accurate delineation of floodwater boundaries.

Aims of this study in relation to flood detection and mapping were to utilise remote sensing methods to (a.) characterise the flood cycles of the lake; (b.) map the spatial distribution of water across the floodplain, and; (c.) determine the relationship between the flooding cycles of the lake and vegetation distributions across the floodplain. The current flood monitoring and mapping efforts of the MNRC and MRC rely on simple linear models of the relationship between river gauge height collected at only a few locations and maximum annual volume and flooded area. Few of the tributaries which drain the 13 catchments around the lake and make significant contributions to lake volume and flooded area have any gauging stations, and hydrological relationships between these tributaries and the lake are complex (Penny, 2006).

For the current study, regional scale MODIS (Moderate resolution Imaging Spectrometer) data was used to determine inundation patterns. MODIS images in 36 spectral bands at 250 m, 500 m and 1 km resolutions, dependent on wavelength, and is widely used for multiple land and ocean applications which require high frequency temporal coverage (Lillesand et al. 2008). A large time-series of MODIS 500m 8 Day Surface Reflectance imagery collected over the period 2001-2005 was used to characterize the flood cycles during the period June to March, at weekly intervals, where the data was of sufficient quality. The MODIS imagery was subsetted to the area of the Tonle Sap and rectified to the ASTER basemap (Figure 2) with its much higher spatial precision using 6 GCPs per image.

The temporal dynamics of the flooded area for the lake are affected by landcover, infiltration rates, and local catchment inputs and cannot be estimated simply from lake gauge height. Inundation mapping in floodplain environments can be problematic due to the presence of high levels of vegetative cover, shallow inundation over large areas and dark organic rich alluvial soils which can appear inundated when they are not (Pearce, 1995). The methods used to map inundation can have a marked effect on the observed patterns (Frazier et al. 2003). On the Tonle Sap, the use of AIRSAR and JERS-1 radar data has been investigated as means of mapping inundation at localised scales (Milne & Tapley, 2005) but this has not been applied at the scale of the entire floodplain. Usual inundation mapping methods using optical imagery involve use of a ratio of mid-infrared reflectance to a visible band reflectance (Lillesand et al. 2008) although this is generally only suitable for relatively deep water. Investigations of techniques for floodplains suggest a combination approach using this ratio and mid-infrared (MIR) change detection is necessary to deal effectively with the shallow water problem (Sims, 2004). Due to the unique nature of the floodplain vegetation and shallow inundation over much of the wet season lake area, a specialised flood detection algorithm was developed for the Tonle Sap using MODIS B6/B4 ratio combined with a B1 threshold, and the accuracy of the technique was verified using the wet-season ASTER imagery. An example of the output from this analysis for a single image date is shown in Figure 6. The MODIS time series was used to determine the extent of flooding and flood duration in conjunction with hydrological data from the CNMC and Mekong River Commission.

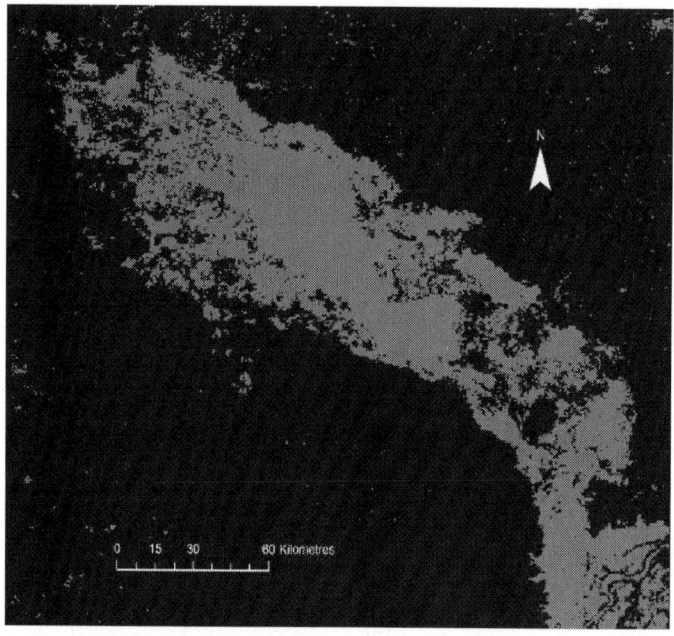

Figure 6. MODIS derived flood extents for the Tonle Sap – an example.

The MODIS derived flood maps indicate a reduction in flooding extent of the Tonle Sap lake since 2000. While the 2000 flood was large by historical standards, and caused widespread damage and loss of life throughout the Mekong Basin (CNMC, 2006), every year since then has been characterised by a reduction in the spatial extent of flooding across the floodplain, apart from the 2008 flood for which the MODIS imagery products are not yet available. This corresponds with MNRC and MRC observations that the flood peaks are now reduced in amplitude and have a much faster fill and drain cycle (CNMC, 2006). Some authors have suggested large dam development throughout the Mekong, and particularly in China, may be responsible (Blake, 2001). The very large Dachaosan dam in southern China began filling in 2003. The monsoons deliver large quantities of water very quickly into the dams where it can be released slowly throughout the year for hydroelectricity generation and for irrigation. The Chinese government currently has another three dams under construction in the upper reaches of the Mekong, with the Xiaowan dam now nearing completion, and another three are at the planning stage (Osbourne, 2006). This will form an 8 dam cascading system capable of retaining very large volumes of water that would otherwise contribute to the monsoonal Mekong flood pulse. With limited fossil fuel reserves and exponential growth in energy demand, the Mekong and other Chinese rivers are seen as offering abundant cheap and clean power. The Chinese dams in the upper reaches of the Mekong are unlikely to be responsible for all reduced flow into the Tonle Sap, as the region may also be experiencing some ongoing effects of drought and climate change (MRC, 2005), and irrigation development is also occurring rapidly on other tributaries which feed the lake. Other current and proposed dams for Laos, Thailand and Vietnam are likely to further ameliorate the Mekong flood pulse in the future. The output from the analysis of the MODIS time-series was then used to model the effects of inundation variability on the wetland and floodplain vegetation on the Tonle Sap floodplain.

GIS MODELLING OF THE EFFECTS OF FLOODING CHANGES ON VEGETATION

The MODIS time-series for the period 2001-2005 shows the area of the Tonle Sap flooded each year and duration of inundation. A goal of the current study was to be able use all the remotely sensed data and derived information on the functioning and spatial arrangement of vegetation and landuse on the floodplain to predict what changes might occur due to interference with the annual flood pulse. This entailed determining the flooding characteristics of floodplain vegetation in terms of depth, timing and duration of flooding and relating these to the spatial distribution of changes in flood patterns. The effects of possible diminished flood peak height and duration on the floodplain were simulated by using an average dry year hydrograph averaged from the four years 1992, 1993, 1999 and 2003 from CNMC data for the Tonle Sap to modify the maximum flood extent model derived from the MODIS imagery. Increasing water use and extraction throughout the Mekong is likely to create move toward dry year conditions with reduced water availability. The average dry year extent was subtracted from the average maximum flood extent derived from the five years of MODIS data for the period 2001-2005, and processed at the resolution of the floodplain dem (30 m). The results show the likely changes in the extent of flooding on the Tonle Sap floodplain if the flooding was likely to be reduced to drier year conditions due to water resource development in the Mekong Basin (Figure 7). When used with the vegetation and landuse cover classifications of the floodplain, this enables GIS modelling of the changes likely to occur in respective landcover types due to reductions in flooded area.

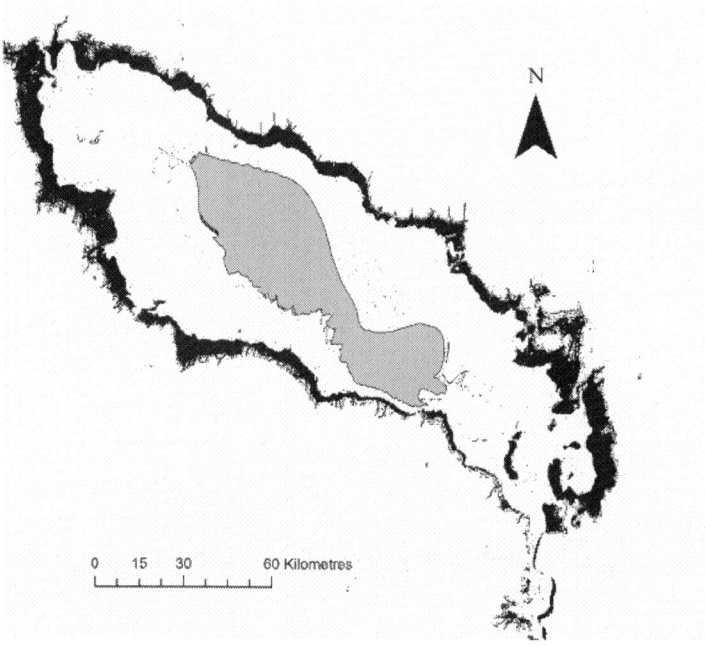

Figure 7. Modelled reduction in flooded area from MODIS derived flood extents, with change shown in black.

The temporal dynamics of the annual flood event on the Tonle Sap are also revealed in the MODIS data, and show the variability in the duration of the flooding. A cell-based GIS analysis was used to calculate the change in the duration of flooding between the simulated dry year conditions and average conditions. This approach also allows for the determination of how changes occur over temporal cycles, making it possible to develop a dynamic means of estimating changes to vegetation and land use types.

The temporal flood cycle model derived from the MODIS time-series and the dem data were integrated into ArcGIS 9.3 Model Builder (Environmental Systems Research Institute, 2009). The time series data provides weekly time steps showing change in flooded area. The mean duration of inundation per cell of land cover type was then generated and this data is summarised in Table 3, showing the change in flood residence time for primary vegetation units between the simulated dry season flood and an average flood. Results indicate that the largest reductions in flood duration will be experienced by the Savannah woodland communities, followed by the shrubland communities, with minor change in the sedge and grassland communities. The core wetland areas of flooded forest and reeds occur at lower elevations and show no reduced flood duration in this analysis. The results of simulating reduced flooding based on the average dry year hydrological data for the Tonle Sap from the GIS-based flood extent model indicate that reductions in flood peak and duration such as those experienced during dry years will have a significant effect on inundation area of the floodplain. This would result in reduction in flooded area reducing from 13,286 km^2 to 11,134 km^2, or approximately 16%.

Table 3. Change in flood durations between simulated dry year and average flood conditions for primary vegetation classes.

Vegetation Class	Simulated Dry Year Flood Duration (days)	Average Flood Duration (days)	Change (days)
Barringtonia acutangula dom.	318	318	0
Barringtonia acutangula dom.	242	311	69
Diospyros cambodiana dom.	213	288	75
Euphorbiaceae Shrubland	58	79	21
Tiliaceae Shrubland	51	68	17
Sedge	43	52	9
Phragmites Reeds	324	324	0
Grassland	11	23	12

Core areas of wetland on the floodplain including the high conservation value *Barringtonia acutangula*dominated flooded forests are most immune to changes in the flood amplitude as they are subject to greater depths of inundation and these are estimated to decline in area by only 2.5%. However, reduced lake levels and reduced flood duration will mean that normal full canopy submergence may no longer occur or submergence time will be reduced. This may affect productivity and growth characteristics and cause a transition towards shorter trees. Similarly, core emergent reed and grass mat areas will suffer only limited effects and as they are short rooted and colonise quickly they can more easily make spatial transitions. Flooded woodland savannah, which makes up the majority of the floodplain is likely to be significantly affected, with areas predicted to reduce by some 23%. Grassland and sedge communities on the distal margins of the floodplain will be greatly reduced in area by an estimated 76%, although they are fast disappearing anyway due to human encroachment. In terms of human landuse, dry season cropping area within the flooded zone will be reduced by an estimated 43%, which will displace these activities to other locations, most likely toward lower elevations in the floodplain. The infrastructure associated with dry season cropping will in many cases no longer be viable.

The results of the GIS modelling indicate that a number of habitats within the Tonle Sap floodplain are vulnerable to changes in the monsoonal flood pulse. This will possibly have ramifications throughout the Mekong due to the importance of many areas as fish breeding habitat. These problems will be compounded by the incursion of agricultural activities into core wetland areas as water availability is reduced on the lake margins (Campbell et al. 2006), along with associated land clearance and resource extraction.

CONCLUSION

Remote sensing is able to provide valuable information on the structure, processes and functioning of the Tonle Sap floodplain. Large, inaccessible wetland and floodplain systems such as the Tonle Sap can be studied from space with a range of remote sensing technologies in combination with appropriate fieldwork and reference data. Interference with the natural flood cycles and inundation patterns of the lake and surrounding floodplain are causing changes in vegetation and are likely to be affecting the biological productivity not only on the Tonle Sap but throughout the Mekong system. The myriad impacts occurring in and around the impoundment structures on the floodplain are changing wetland community composition and structure, which in turn will affect fisheries productivity and species biodiversity. Local livelihoods are already affected by fierce (often violent) competition for lake and floodplain resources (Bonheur & Lane, 2002), and as the wetlands and floodplain degrade further this is likely to increase. Historical development of water resources has had significant impacts on the environments and catchments in parts of the floodplain, and caused permanent changes in the hydrology of these areas (Kummu, 2009), and this will continue and accelerate with population growth in the region. Water resource use upstream of the Tonle Sap is potentially reducing and moderating the monsoonal flood pulse which sustains the lake and floodplain system. This may be linked to the timing of large dam construction within the Mekong River basin, although Laos and Thailand are extracting increasing amounts of water from the Mekong as well for use in rapidly expanding rice irrigation schemes (Osbourne, 2006). While social benefits may arise from amelioration of floods which in some years can cause extensive property damage and loss of life, this must be balanced against the need to maintain flood cycles which can sustain the environment of the Tonle Sap, and economic activities such as fishing and agriculture. There is an urgent need to develop effective cross-border management plans and agreements for the water resources of the Mekong system before the unique and economically important Tonle Sap region slips into further decline.

Future events in the Mekong basin, whether related to climate change or human development, will have important ramifications for the Tonle Sap. The annual flood pulse which sustains lake and floodplain ecology is vulnerable to change and as it changes the primary vegetation communities on the Tonle Sap floodplain will most likely face significant declines. In addition, in-situ impacts from upstream developments in the sub-catchments of the lake, as well as further modification of the floodplain will act to reduce water availability and wetland area. The floodplain is already exhibiting signs of over-exploitation (Campbell et al. 2006) and this will increase in line with population and development pressures. It is critical that future basin planning and water resource extraction between the Mekong Basin countries be coordinated in order to preserve the size, duration and timing of the flooding of the Tonle Sap.

REFERENCES

1. Asian Development Bank 2002 Report and Recommendation for the Tonle Sap Environmental Management Project, ADB Report RRP: Cam 33418.
2. Asian Development Bank 2005 Summary Initial Environmental Examination Report for the Tonle Sap Sustainable Livelihoods Project in Cambodia, August 2005, ABD.
3. Asian Development Bank 2009 The Tonle Sap Initiative: Future Solutions Now, Available online:http://www.adb.org/Projects/Tonle_Sap/default.asp
4. S. N. Benger, 2006 Groundwater interactions with the wetlands of the Tonle Sap, Cambodia, in Proc. HydroEco 2006, Karlovy Vary, Czech Republic, Sept 2006, 45 48 . 8-09036-351-2
5. D. Blake, 2001 Proposed Mekong Dam scheme in China threatens millions in downstream countries. World Rivers Review 4-5, 43 51, 08906211
6. F. Blasco, M. F. Bellan, M. U. Chaudhury, 1992 Estimating the extent of floods in Bangladesh using SPOT data, Remote Sensing of Environment 39, 167 178 , 0034-4257
7. N. Bonheur, B. D. Lane, 2002 Natural resources management for human security in Cambodia's Tonle Sap Biosphere Reserve, Environmental Science and Policy 51, 33 42 , 1462-9011
8. P. V. Bolstad, T. M. Lillesand, 1991 Rapid maximum likelihood classification, Photogramm. Eng. Remote Sens. 57, 67 74, 0034-4257
9. P. Brunner, H. J. Hendricks Franssen, L. Kgotlhang, P. Bauer-Gottwein, W. Kinzelbach, 2007 How can remote sensing contribute to groundwater modelling? Hydrogeology Journal 15(1), 5 18, 1431-2174
10. Cambodia National Mekong Committee (CNMC) 2006 Cambodia Country Report: Flood information in Cambodia, Proceedings of the 4th Annual Mekong Flood Forum "Improving Flood Forecasting and Warning Systems for Flood Management and Mitigation in the Lower Mekong Basin", Siem Reap, Cambodia, May 2006, 23 36 .
11. I. C. Campbell, C. Poole, W. Giesen, J. Valbo-Jorgensen, 2006 Species diversity and ecology of Tonle Sap Great Lake, Cambodia, Aquatic Sciences-Research Across Boundaries 68, 355 373, 1015-1621
12. D. Chandler, 1996 A History of Cambodia, Westview Press Inc, 9-74710-065-7

13. B. J. Choudery, 1991 Passive microwave remote sensing contribution to hydrological variables, Surveys in Geophyics 12, 63 84 , 0169-3298

14. J. B. Collins, C. E. Woodcock, 1996 An assessment of several linear change detection techniques for mapping forest mortality using multitemporal Landsat TM data. Remote Sensing of Environment 56 66 77 , 0034-4257

15. R. G. Congalton, K. Green, 2008 Assessing the Accuracy of Remotely Sensed Data: Principles and Practices, 2nd Edition, CRC Press, 978-1-42005-512-2 Boca Raton, FL.

16. H. Daming, 1997 Facilitating regional sustainable development through integrated multi-objective utilization management of water resources in the Lancang-Mekong river basin, Journal of Chinese Geography 7, 4, 1861-9568

17. N. Engheta, C. Elachi, 1982 Radar scattering from a diffuse vegetation layer over a smooth surface, IEEE Trans. Geosci. and Remote Sens. 20, 212 216 , 0196-2892

18. Environmental Systems Research Institute 2009 ArcGIS 9.3 ModelBuilder Software, ESRI, Redlands, CA.

19. D. Evans, C. Pottier, R. Fletcher, S. Hensley, I. Tapley, A. Milne, M. Barbetti, 2007 A comprehensive archaeological map of the world's largest preindustrial settlement complex at Angkor, Cambodia, Proceedings of the National Academy of Sciences of the United States of America 104 (36), 14277 14282 , 1091-6490

20. D. E. Evans, T. G. Farr, J. P. Ford, T. W. Thompson, C. L. Werner, 1986 Multipolarisation radar images for geological mapping and vegetation discrimination, IEEE Trans. Geosci. and Remote Sens. 24, 246 257 , 0196-2892

21. J. P. Ford, D. J. Casey, 1988 Shuttle radar mapping with diverse incidence angles in the rainforest of Borneo, Int. Journal of Remote Sensing 5, 927 943 , 1366-5901

22. J. P. Ford, J. B. Cimano, B. Holt, M. R. Ruzek, 1986 Shuttle Imaging Radar Views the Earth from Challenger: The SIR-B Experiment, Jet Propulsion Laboratory publication 86-10, Pasadena, California

23. R. T. T. Forman, M. Godron, 1986 Landscape Ecology, Wiley, 0-47187-037-4 New York.

24. P. Frazier, K. Page, J. Louis, S. Briggs, A. I. Robertson, 2003 Relating wetland inundation to river flow using Landsat TM data. Int. Journal of Remote Sensing 2419, 3755 3770 , 1366-5901

25. H. J. Hendricks Franssen, P. Brunner, P. Makobo, W. Kinzelbach, 2008 Equally likely inverse solutions to a groundwater flow problem including pattern information from remote sensing images, Water Resources Research, 44, W01419, doi:10.1029/2007WR006097

26. L. L. Hess, J. M. Melack, D. S. Simonett, 1990 Radar detection of flooding beneath the forest canopy: a review, Int. Journal of Remote Sensing 11, 1313 1325 , 1366-5901

27. C. Higham, 2001 The Civilisation of Angkor, Orion Books, London. 1-84212-584-2

28. K. Hixson, D. Scholz, N. Funs, 1980 Evaluation of several schemes for classification of remotely sensed data, Photogramm. Eng. Remote Sens. 46, 1547 1553 , 0099-1112

29. R. M. Hoffer, D. F. Lozano-Garcia, D. D. Gillespie, P. W. Mueller, M. J. Ruzek, 1986 Analysis of multiple incidence angle SIR-B data for determining forest stand characteristics, The Second Spaceborne Imaging Radar Symposium, JPL Publication 86-26, Pasadena CA, 159 164

30. F. M. R. Hughes, 1990 The influence of flooding regimes on forest distribution and composition in the Tana River Floodplain, Kenya, Journal of Applied Ecology 27, 475 491 , 1365-2664

31. V. Ingegnoli, 2002 Landscape Ecology: A Widening Foundation, Springer, 978-3-54042-743-8 Amsterdam.

32. M. Imhoff, C. Vermillion, M. H. Story, A. M. Choudery, A. Gafoor, 1987 Monsoon flood boundary delineation and assessment using spaceborne imaging radar and Landsat data, Photogramm. Eng. Remote Sens. 53, 405 413 , 0099-1112

33. J. W. Jacobs, 2002 The Mekong River Commission: transboundary water resources planning and regional security, The Geographical Journal 168, 354 364 , 1861-9568

34. J. R. Jensen, 2007 Remote Sensing of the Environment, 2nd Edition, Pearson Prentice Hall, 0-13188-950-8 Upper Saddle River, NJ.

35. S. K. Jensen, F. A. Waltz, 1979 Principal Components Analysis and Canonical Analysis in Remote Sensing, in Proc. American Photogrammetric Society 45th Annual Meeting, 337 348 , 0099-1112

36. R. M. Johnson, M. M. Barson, 1993 Remote sensing of Australian wetlands: an evaluation of Landsat TM data for inventory and classification, Aust. J. Mar. Freshwater Res. 44, 235 252 , 0067-1940

37. W. J. Junk, P. B. Bayley, R. E. Sparks, 1989 The Flood Pulse Concept in river-floodplain systems, Can. Spec. Publ. Fish. Aquat. Sci. 106, 110 127 . 1205-7533

38. R. J. Kauth, G. S. Thomas, 1976 The Tasseled Cap- A graphic description of the spectral temporal development of agricultural crops as seen by Landsat, in Proc. LARS 1976 Symposium on Machine Processing of Remotely Sensed Data, Purdue University.

39. A. S. Kiem, H. Ishidaira, H. P. Hapuarachchi, M. C. Zhou, Y. Hirabayashi, K. Takeuchi, 2008 Future hydroclimatology of the Mekong River basin simulated using the high-resolution Japan Meteorological Agency (JMA) AGCM, Hydrological Processes 22 1382 1394 , 1099-1085

40. A. S. Kiem, H. P. Hapuarachchi, K. Takeuchi, 2004 Impacts of climate variability on streamflow in the Mekong River: an interesting challenge for hydrological modelling, In: Proc. River Symposium 2004, Sept 2004, Brisbane.

41. B. Kiernan, 1996 The Pol Pot Regime- Race, Power and Genocide in Cambodia under the Khmer Rouge, 1975-79, Yale University Press, 9-74710-043-6 New Haven

42. G. Kite, 2001 Modelling the Mekong: hydrological simulation for environmental impact studies, Journal of Hydrology 253, 1 13 , 0022-1694

43. J. Koponen, J. Josza, H. Lauri, J. Sarkkula, V. Markku, 2003 Modelling Tonle Sap Watershed and Lake Processes for Environmental Change Assessment, Mekong River Commission MRCS/WUP-FIN Model Report.

44. M. D. Krohn, N. M. Milton, D. B. Segal, 1983 Seasat synthetic aperture radar (SAR) response to lowland vegetation types in eastern Maryland and Virginia, J. Geophys. Res. 88, 1937 1952 , 0148-0227

45. M. Kummu, 2009 Water management in Angkor: Human impacts on hydrology and sediment transportation, Journal of Environmental Management 90, 3, 1413 1421, 0301-4797

46. T. M. Lillesand, R. W. Kiefer, J. W. Chipman, 2008 Principles of Remote Sensing and Image Analysis, 6th Edition, Wiley, 978-0-47005-245-7 New York.

47. P. A. Longley, M. F. Goodchild, D. J. Maguire, D. W. Rhind, 2005 Geographic Information Systems and Science, 2nd Edition, Wiley, 0-47087-000-1 New York.

48. G. P. Malanson, 1993 Riparian Landscapes, Cambridge University Press, 9780521384315, Cambridge

49. F. Maselli, C. Conese, G. Zipoli, M. A. Pittau, 1990 Use of error probabilities to improve area estimates based on maximum likelihood classifications, Remote Sensing of Environment 31, 155 160 , 0034-4257

50. J. McDonald, P. Bunnat, P. Virak, 1997 Plant Communities of the Tonle Sap Floodplain, UNESCO/IUCN/WI, Phnom Penh.

51. Mekong River Commission 2007 Annual Mekong Flood Report 2006, Mekong River Commission, 1728-3248 Vientiane.

52. Mekong River Commission 2005 Overview of the Hydrology of the Mekong Basin, Mekong River Commission, 1728-3248 Vientiane, November 2005.

53. Mekong Secretariat 1994 Annual Report 1994, Mekong Secretariat, 1728-3248 Bangkok.

54. L. A. K. Mertes, D. L. Daniel, J. M. Melack, B. Nelson, L. A. Martinelli, B. R. Forsberg, 1995 Spatial patterns of hydrology, geomorphology, and vegetation on the floodplain of the Amazon River in Brazil from a remote sensing perspective, Geomorphology 13, 215 232 , 0169-555X

55. L. A. K. Mertes, M. O. Smith, J. B. Adams, 1993 Estimating suspended sediment concentrations in surface waters of the Amazon River wetlands from Landsat images, Remote Sensing of Environment 43, 281 301, 0034-4257

56. A. K. Milne, I. J. Tapley, 2005 Change Detection Analysis in the Wetlands Using JERS-1 Radar Data: Tone Sap Great Lake, Cambodia, IEEE doi 0-7803-9119-5/05. 146 150

57. C. Milzow, N. Kgotlhang, P. Bauer-Gottwein, P. Meier, W. Kinzelbach, 2009 Regional review: the hydrology of the Okavango Delta, Botswana- processes, data and modelling, Hydrogeology Journal, 1431-2174 Published Online DOI 10.1007/s10040-009-0436-0

58. S. Mutiti, J. Levy, C. Mututi, N. S. Guturu, 2008 Assessing Ground Water Development Potential Using Landsat Imagery, Groundwater, Published Online DOI 10.1111/j.1745-6584.2008.00524.x

59. J. P. Ormsby, B. J. Blanchard, A. J. Blanchard, 1985 Detection of lowland flooding using active microwave systems, Photogramm. Eng. Remote Sens. 51, 317 328 , 0099-1112

60. M. Osbourne, 2006 River at risk: The Mekong and the water politics of China and Southeast Asia, Lowy Institute for International Policy, 1-92100-402-9 New York.

61. B. Pearce, 1995 The compilation of regional flood maps using remote sensing techniques over the Ballonne river catchment and downstream areas. Technical Report. Queensland Department of Primary Industries, Brisbane, QLD.

62. D. Penny, 2006 The Holocene history and development of the Tonle Sap, Cambodia. Quaternary Science Reviews 25, 310 322 , 0277-3791

63. K. O. Pope, E. J. Sheffner, K. J. Linthicum, C. L. Bailey, T. M. Logan, E. S. Kasischke, K. Birney, A. R. Nlogu, C. R. Roberts, 1992 Identification of the central Kenyan Rift Valley fever virus vector habitats with Landsat TM and evaluation of their flooding status with airborne imaging radar, Remote Sensing of Environment 40, 185 196 , 0034-4257

64. L. Puy, S. Lek, S. T. Touch, S-O. Mao, B. Chhouk, 1999 Diversity and spatial distribution of freshwater fish in Great Lake and Tonle Sap river Cambodia, Southeast Asia, Aquatic Living Resources 126, 379 386 , 0990-7440

65. S. R. Ramireddygari, M. A. Sophocleous, J. K. Koelliker, S. P. Perkins, R. S. Govindaraju, 2000 Development and application of a comprehensive simulation model to evaluate impacts of watershed structures and irrigation water use on streamflow and groundwater: the case of Wet Walnut Creek Watershed, Kansas, USA. Journal of Hydrology 2363-4, 223 246 , 0022-1694

66. J. A. Richards, X. Jia, 2006 Remote Sensing Digital Image Analysis- An Introduction, 4th Edition, Springer-Verlag, Berlin. 978-3-54025-128-6

67. J. A. Richards, G-Q. Sun, D. S. Simonett, 1987a L-band radar backscatter modelling of forest stands, IEEE Trans. Geosc. Remote Sens. 25, 487 498 , 0196-2892

68. J. A. Richards, P. W. Woodgate, A. K. Skidmore, 1987b An explanation of enhanced radar backscattering from flooded forests, Int. Journal of Remote Sensing 8, 1093 1100 , 1366-5901

69. J. San Miguel Ayanz, G. S. Biging, 1997 Comparison of single-stage and multi-stage classification approaches for cover type mapping with TM and SPOT data, Remote Sensing of Environment 59, 92 104 , 0034-4257

70. M. Scheffer, 1998 The Ecology of Shallow Lakes, Chapman and Hill, 0-41274-920-3, London

71. I. Scoones, 1991 Wetlands in Drylands: key resources for agricultural and pastoral production in Africa, Ambio 20, 366 371 , 0044-7447

72. N. Sims, 2004 The Landscape-scale Structure and Functioning of Floodplains, Unpublished PhD Thesis, University of Canberra.

73. S. J. Sippel, S. K. Hamilton, J. M. Melack, B. J. Choudery, 1994 Determination of inundation area in the Amazon River floodplain using the SMMR 37 GHz polarisation difference, Remote Sensing of Environment 48, 70 76 , 0034-4257

74. J. A. Slater, G. Garvey, C. Johnston, J. Haase, B. Heady, G. Kroenung, J. Little, 2006 The SRTM data "finishing" process and products. Photogramm. Eng. Remote Sens. 72(3), 237 247 , 0099-1112

75. P. Someth, N. Kubo, H. Tanji, S. Lyd, 2009 Ring dike system to harness floodwater from the Mekong River for paddy rice cultivation in the Tonle Sap Lake floodplain in Cambodia, Agricultural Water Management 96, 100 110 , 0378-3774

76. G. Stanger, T. Van Truong, K. S. Ngoc, T. V. Luyen, T. T. Thanh, 2005 Arsenic in groundwaters of the Lower Mekong, Environmental Geochemistry and Health 27, 341 357 , 1573-2983

77. N. Top, N. Mizoue, S. Kai, T. Nokao, 2004 Variation in woodfuel consumption patterns in response to forest availability in Kampong Thom Province, Cambodia, Biomass and Energy 27, 57 68 , 0167-5494

78. N. Van Zalinge, N. Thouk, T. C. Tana, D. Leung, 2000 Where there is water, there is fish? Cambodian fisheries issues in a Mekong River Basin perspective. In: Ahmed, M. and Hirsh, P. (Eds) Common Property in the Mekong: Issues of Sustainability and Subsistence. ICLARM Study Review.

79. R. Webby, P. T. Adamson, J. Boland, P. G. Howlett, A. V. Metcalfe, J. Piantadosi, 2005 The Mekong- Applications of Value at Risk (VaR) and Conditional Value at Risk (CVaR) simulation to the benefits, costs and consequences of water resources development in a large river basin. In: MODSIM 2005 International Congress on Modelling and Simulation. (ed. by A. Zerger & R.M. Argent), 2109 2115, 0-97584-000-2, Modelling and Simulation Society of Australia and New Zealand, December 2005, Brisbane.

80. E. Wikramanayake, E. Dinerstein, 2001 Terrestrial Ecoregions of the Indo-Pacific, Island Press, 1-55963-923-7 Washington DC.

81. G. Wright, D. Moffatt, J. Wager, 2004 Establishment of the Tonle Sap Basin Management Organisation: Tonle Sap Basin Profile, Cambodia National Mekong Committee, Asian Development Bank Report TA2412-CAM.

82. S. T. Wu, S. A. Sadler, 1987 Multipolaristaion SAR data for surface feature delineation and forest vegetation characterisation, IEEE Trans. Geosc. Remote Sens. 25, 67 76 , 0196-2892

83. S. R. Yool, Y. L. Star, J. E. Estes, E. B. Botkin, D. W. Eckardt, F. W. Davis, 1986 Performance analysis of image processing algorithms for classification of natural vegetation in the mountains of southern California, Int. Journal of Remote Sensing 7, 683 702 , 1366-5901

CITATION

Simon Nicholas Benger (2009). Remote Sensing of the Ecology and Functioning of the Mekong River Basin with Special Reference to the Tonle Sap, Geoscience and Remote Sensing, Pei-Gee Peter Ho (Ed.), ISBN: 978-953-307-003-2, InTech, DOI: 10.5772/8296.

CHAPTER 6

Surveying and Monitoring for Vulnerability Assessment of an Ancient Building

Luigi Fregonese, Gaia Barbieri , Luigi Biolzi , Massimiliano Bocciarelli, Aronne Frigeri and Laura Taffurelli

Department of Architecture, Built Environment and Construction Engineering, ABC, Politecnico di Milano, via Ponzio, Milano 31-20133, Italy

ABSTRACT

This paper examines how surveying and monitoring improve our knowledge about ancient buildings, allow the interpretation of their structural response and help in the search for the best solutions for their conservation. The case study of *Palazzo del Capitano* in Mantua (Italy) is analyzed. In particular, the attention is focused on the use of a Terrestrial Laser Scanner (TLS) for surveying and monitoring too, considering that the building structural control has been performed in combination with other traditional topographic techniques such as geometric leveling and topographic networks for 3D control based on measurements through total stations. The study of TLS monitoring has been tested only in the last decade and it is an innovative method for the detection of displacements of particular surfaces. Till now the research has focused only on the use of TLS monitoring to control large structures and in particular landscape situations; thus its use for a civil construction and historical buildings is a new field of investigation. Despite the fact technological development and new methodologies seem offer new future potential for the analysis of ancient buildings, currently there are still important limits for the application of the investigated surveying and monitoring techniques.

INTRODUCTION

The increasing importance that current societies give to their historical and cultural heritage has promoted the drafting of specific legislations aimed at protecting and preserving it. In particular, Italian Codes [1,2] assert that the topographic techniques and their instruments provide an essential support to assess the structural response of an ancient building, for a conscious conservation. These Codes underline:

- the importance of having a precise geometrical survey in order to know the building in details. This is used in order to detect the anomalies and deformations of the buildings surfaces and, as a consequence, to set up an highly realistic structural 3D model for a Finite Elements Method (FEM) analysis;
- the importance of planning an appropriate monitoring program, based on the identification and interpretation of the buildings deficiencies. In this way it is possible both to control some significant parameters of the structures in time (crack movements, absolute or relative displacements of some points, rotation of walls or other elements) and to have a better understanding of their global structural behavior.

In the current paper, the contribution of the topographic techniques and instruments to the structural analysis of the cultural heritage was considered in relation to *Palazzo del Capitano*, one of the main masonry buildings of Mantua, Italy, see Figure 1.

Figure 1. *Palazzo del Capitano* in Mantua. View from *Piazza Sordello*.

This building was built between the last years of the thirteenth century and the first decade of the fourteenth century by the first Lords of Mantua. It is one of the most ancient parts of the *Palazzo Ducale* complex and over the centuries it has undergone a series of transformations which have partially changed its original structure. Since the eighteenth century, documents have shown the presence of significant out of plane displacements characterizing the two longitudinal façades overlooking *Piazza Sordello* and *Piazza Pallone*: they both present an inclination in the direction of *Piazza Pallone*. These static problems are probably due to ground settlement and they are related to the configuration of the structure and its transformations over time.

In the last two years a detailed structural analysis of the building—supported by a programmed monitoring campaign of the possible kinematic mechanisms that are still ongoing—has been conducted. This work was planned in order to analyze the structural response of the building and also to suggest a possible restoration.

The second section of the current paper deals with the geometrical survey in view of the structural analysis of *Palazzo del Capitano*. Before the structural analysis, the survey of the building was performed and completed through the integration of different techniques. The technology development in time allowed us firstly to use a direct and photogrammetric survey and secondly a laser scanner one. Thanks to this tool, the building was digitally acquired by means of dense point clouds which detected accurately the anomalies and deformations of the masonry walls. A 3D numerical model obtained was then used to perform the structural analysis. These allowed us to examine the building behavior both under serviceability conditions and in respect of exceptional events expected by current codes (such as an earthquake).

However, nowadays the potentialities opened by the TLS technique in relation to a new approach to the structural analysis have considerable limits. These are due to both the operative problems in large-size object acquisition and in data processing, and moreover to the transposition of the survey information in a numerical 3D model.

The third section of the paper deals with all the features related to the monitoring of *Palazzo del Capitano*. Simultaneously to the geometrical survey and structural analysis, the monitoring of the building was conducted for about one year with monthly or multimonthly cadences.

Monitoring was aimed mainly at researching possible ground sinking or ongoing increases in the permanent out of plane displacements of the front façade overlooking *Piazza Sordello*. The periodic control of the movements in time allowed us to verify firstly the building structural behavior under normal operating conditions and secondly also under particular conditions that might have occurred during the period of observation.

In particular, in this paper, the attention has been focused to the use of the TLS for the control of the surfaces movements. Even in this case, the potential related to the use of such technique—which from a theoretic point of view allowed us a wide and very accurate control on a large number of points of the object under consideration—has clashed with the limits that are mainly related to problems of stability and precision of the georeferencing scans.

GEOMETRICAL SURVEY FOR STRUCTURAL MODELING

The starting point for the knowledge of historical buildings belonging to the cultural heritage is the geometrical survey. Clearly, it must be supported by an in-depth historical and archival research and by thorough surveys on materials, structural techniques and geological aspects of the building site [1,2].

The integration of such surveys is essential to examine the buildings which, generally built before the twentieth century, are lacking in documentations about their construction, evolution and conservation. Most of these buildings were built when the common approach to design and construction was based on experience and empirical rules. Moreover, workers and available techniques in the past couldn't control the realization of certain structures (especially of large sizes) with the precision that we can get nowadays. Often, time and historical events have damaged ancient buildings, with significant distresses, so that sometimes the buildings have changed or even distorted their originally structures.

In order to have a complete and thorough knowledge of the historical buildings it is essential to consider all the cognitive contributions provided by the geometrical survey. Indeed, it has to provide both the documentation and representation for the artistic and architectural aspects and the metric information for the structural analysis.

In particular, the ongoing development of the laser scanning technology allows us to reproduce accurately the three-dimensional geometry of all buildings acquired. In relation to the structural analysis, the TLS data can be used to reconstruct actual deformed configuration of the entire masonry surfaces of the buildings directly [3–5]. It can also used to provide the information needed for the elaboration of 3D models which allow us to investigate the expected behavior of the structure in particular conditions by means of FEM analyses [6–9].

Geometrical Survey of Palazzo del Capitano for the Structural Analysis: From Direct Survey to TLS

Palazzo del Capitano in Mantua rises three floors above ground, up to a height of about 22 m at the eaves line and 24.5 m at the ridge line of the gable roof. It has a rectangular plan of about 67–68.5 m × 16.5 m. Thicknesses of the bearing walls vary between 100 cm and 80 cm on the ground and at first floor, and between 80 cm and 70 cm on the second floor.

Of the two longitudinal façades, the north-west one overlooks *Piazza Sordello*—where there is the "public" side of the building—and the south-east one overlooks *Piazza Pallone*—where there is the "private" side, once overlooking the ancient *brolo*.

The ground floor is characterized by a majestic portico on the *Piazza Sordello* side, and by a succession of rooms on the *Piazza Pallone* side. On the first floor, over the portico, there is a long hallway, known as *Galleria del Passerino*, while on the side overlooking *Piazza Pallone* there are the rooms of the Duke of Guastalla's apartment. The second floor is entirely taken up by the wide hall of the Armoury, known as *Salone dell'Armeria*, which is interposed by masonry partitions built at the beginning of the twentieth century to address the inclination of the longitudinal walls. The plans, façades and sections of the building in its current configuration are illustrated in Figure 2, Figure 3 and Figure 4.

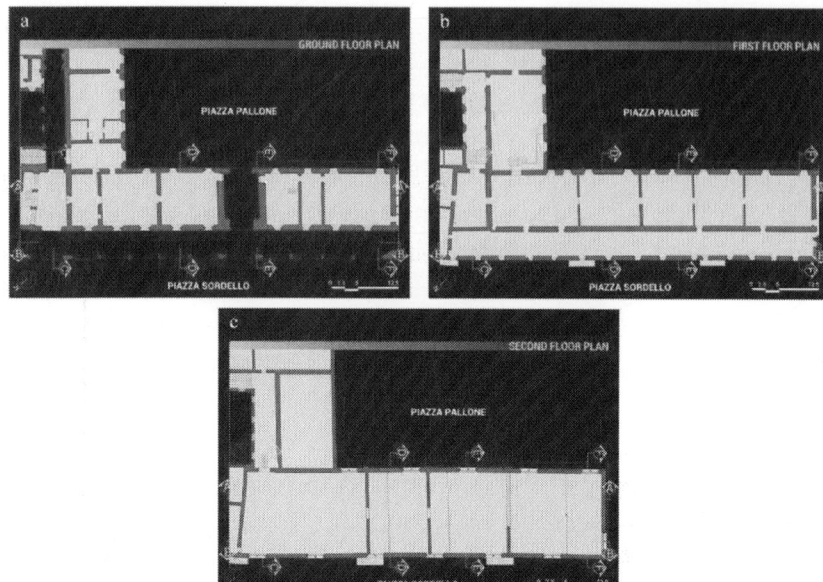

Figure 2. *Palazzo del Capitano* plans (**a**) Ground floor plan. (**b**) First floor plan. (**c**) Second floor plan.

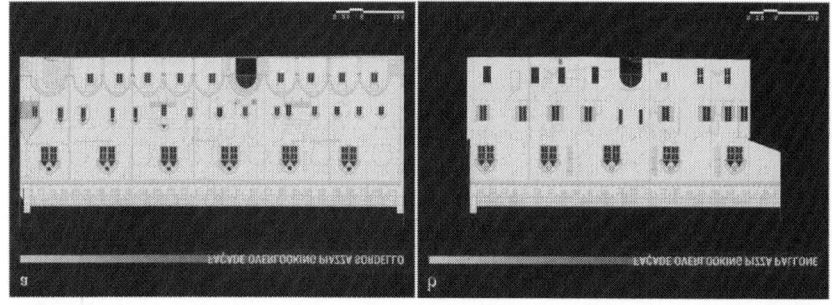

Figure 3. *Palazzo del Capitano* façades (**a**) façade overlooking*Piazza Sordello*. (**b**) façade overlooking *Piazza Pallone*.

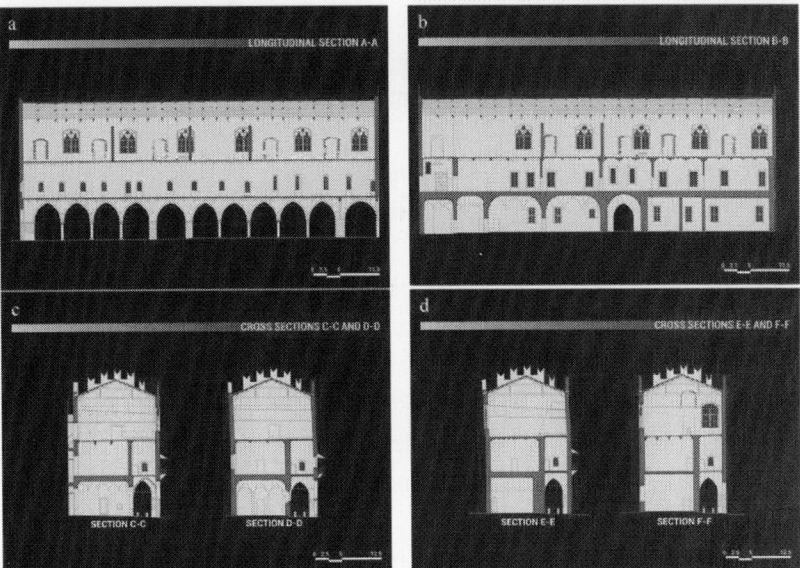

Figure 4. *Palazzo del Capitano* sections (**a**) Longitudinal section AA. (**b**) Longitudinal section BB. (**c**) Cross sections CC and DD. (**d**) Cross sections EE and FF.

The first surveys of *Palazzo del Capitano* were performed in the late sixteenth century, when architects of the palatine court drew up the plans of the building. However, since the early nineties of the twentieth century, the survey of the entire building in its current state has been performed using modern techniques and PC support, with different methods as required by the local superintendence for the historical and architectural heritage.

In particular:

- In 1993 a photogrammetric survey of the building façades and a direct survey of the internal rooms were performed. The plans of all floors, the two fronts and a cross section were executed at a nominal 1:100 scale;
- In 2005 a photogrammetric survey of the building façades and the internal walls of the *salone dell'armeria* was performed. Vettorial restitutions and digital orthophotos at a nominal 1:50 scale were executed. The survey was performed by means of a rollei 6006

camera with a 16 mb phase one digital back. The photogrammetric takes were carried out by a 40 mm lens at a distance of approximately 10 m. The images had a standard deviation σ in acquisition phase of about 2.5 mm, which is lower than the graphical error e_{gr} allowed for a 1:50 nominal scale (e_{gr} = 0.2 × 50 = 10 mm). The orthophotos were made in pci geomatics software and an average standardized residual rsa in a range of 2–3 mm in correspondence of the control points was obtained: this was compatible with the accuracy provided by the support topographic networks;

- Between 2005 and 2007 a laser scanner survey of the building façades, the interior space of the *salone dell'armeria* and the rooms of the duke of guastalla's apartment at the first floor was performed, see figure 5. The survey was conducted using leica hds3000 and leica hds6000 laser scanners, especially to investigate the magnitude of the wall inclinations in a very detailed way;

Figure 5. *Palazzo del Capitano* point cloud in Cyclone.

- Between 2011 and 2012 direct measurements were performed to complete the architectonic survey of the ground floor rooms and of the *Galleria del Passerino* at the first floor plan. The survey consisted of concatenated trilaterated quadrangles (six planimetric observations), framed by topographic control points. The relative tolerance error T of the 1:50 scale was assumed equal to 2.5 times the E_{GR} (T = ±25mm).

The integration of the surveying methods at different accuracy level was performed considering the TLS data as the primary source of the survey. The missing parts of the TLS data were ultimately obtained from the photogrammetric survey and direct measurements. However, thanks to the point cloud georeferencing, the future digital acquisition of the parts of the building not yet examined with the laser technique will allow us to further refine the geometrical survey.

TLS data were processed in Cyclone and then in CAD with the CloudWorks application. In Cyclone the scans were filtered and registered in a georeferenced system. In CAD they were supplemented with the direct and photogrammetric survey and then they were elaborated to evaluate the entity of the structural distresses along the entire surface of the building.

The outcomes of the geometrical survey show clearly the permanent deformation state of the longitudinal façades. The inclination begins in the upper part of the first floor and it increases more or less linearly as a function of the height. This state of deformation is probably due to ground settlement (and non uniform distribution of the soil properties) and to the lack of constraints joining the longitudinal façades to transversal walls. In fact, the wooden floors don't establish a mechanical constraint in the plane and on the second floor of the *Salone dell'Armeria*, the tie-rods and bracing walls (which aren't toothed to the perimetral walls) were inserted only after the detection of the permanent displacements.

TLS data show that the main deformation occurs in the central part at the roof level. The façade overlooking *Piazza Sordello* presents an out of plane displacement ranging from 0.36 m to 0.38 m (from 4.4% to 4.6% of the last storey height), while the other façade overlooking *Piazza Pallone* presents an out of plane displacement from 0.32 m to 0.40 m (from 3.9% to 4.9% of the last storey height), see Figure 6. Finally, an inclination of the floor structure at the level of the *Salone dell'Armeria* of about 0.8% from*Piazza Sordello* (upper level) to *Piazza Pallone* (lower level) has been measured.

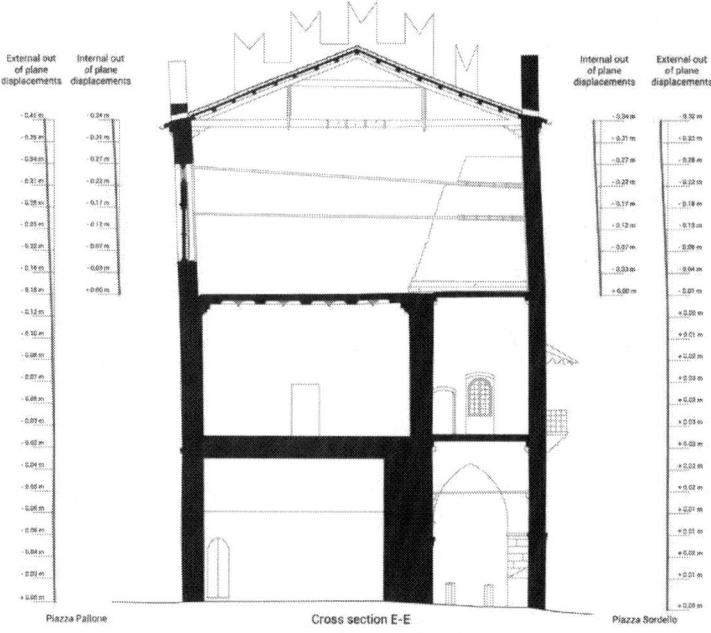

Figure 6. Cross section EE with the indication of the out of plane displacements of the walls as result by the Laser Scanner acquisition.

From some sections made at different heights of the building, a numerical model was realized in 3D-CAD in order to reproduce the actual geometrical configuration in detail, despite the simplifications adopted for some architectonical elements.

FEM Analysis of Palazzo del Capitano

The 3D-CAD geometrical model of the existing deformed configuration, including also the permanent out of plane displacements, was imported into the finite element program Abaqus [10], see Figure 7, and then discretized by four nodes tetrahedral elements for a total number of about 250,000 degrees of freedom. According to experimental measurements at various points of the masonry structure by means of double flat jacks, the average elastic modulus of the ancient masonry was 1,640 MPa. The compressive strength of the material was assumed equal to 2.4 MPa as suggested by the Italian Code [1] in Annex C8A.2.

Figure 7. Geometrical model of the building in its current state. Lines represent the steel tie-rods.

This model was adopted to assess the seismic vulnerability of the structure [11]. In Figure 8 the vertical stress component due to the self weight of the structure are shown. The structure is compressed almost everywhere and only in a few points the analysis detects the presence of tensile stresses, which however are small enough not to lead to the partition of the section. The stress field due to the vertical loads is significant in some points: the highest compressive stress in the masonry walls is registered on the front of *Piazza Sordello*, near the springers of the arches, due to the strong reduction of the resistant section.

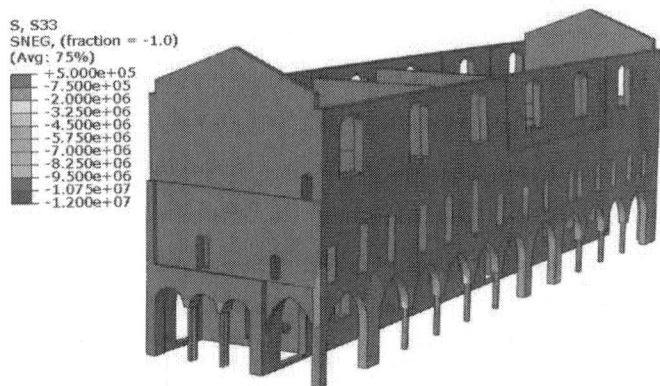

Figure 8. Vertical stress state (expressed in [N/m²]) in the longitudinal façade induced by the self weight of the structure.

Limits and Potentialities of the TLS Geometrical Survey for the Structural Analysis of an Historical Building

From a theoretical point of view, the TLS geometrical survey has great potential, mainly thanks to the accuracy and speed in the data acquisition. The development of the TLS technique allows us to pass from a two-dimensional to a three-dimensional knowledge of the building under investigation. At the beginning, the building is detected only in correspondence of a few sections extracted by the direct measurements of a few remarkable points, later it is digitally acquired by a dense point cloud which describes the structure geometry continuously, including all the irregularities and anomalies of the surfaces.

These potentialities are important even when the geometrical survey is adopted for structural analysis purposes of the historical buildings. At the same level of knowledge of other factors, if the survey accuracy increases (and then it is possible a more accurate three-dimensional model), it is possible to obtain more reliable outcomes from the structural analysis.

However, the use of a laser scanner survey is related to a series of limits and problems that can't be underestimated.

First of all, it is necessary to consider the difficulties normally related to the digital acquisition of an entire building (especially if it has important dimensions, such as *Palazzo del Capitano*) and to the data elaboration in order to obtain a numerical 3D model. In particular:

- the presence of obstacles interfering with digital acquisition of the building—such as vegetation, furniture or people walking while the laser scanner is operating;
- instrument capability, related mainly to their range;
- data processing time in particular related to the point cloud filtering operations. In addition, with large-size building several scans to acquire the object as a whole are required.

Secondly, the limits of the structural software referring to the laser scanner use for a numerical model construction have to be considered. Indeed, a detailed 3D model of every architectural feature, should produce significant calculation complexities and unacceptable data processing time. As a consequence, it is necessary that the point cloud be transferred into a simplified model neglecting the secondary elements which don't participate—or take part only minimally—in the overall structural behavior of the building.

STRUCTURAL MONITORING

Monitoring plays a very important role in the diagnostic process of an historical building. Through monitoring instruments, the evolution (as a function of time and of environmental condition changes) of parameters that could be dangerous for the structural efficiency is controlled. In order to estimate the presence of ongoing permanent kinematic mechanisms, due to intrinsic factors and not to cyclical seasonal phenomena, it is important to establish a suitable period of observation, from 12 to 24 months. Moreover, by means of a trend analysis, the long-term behavior of the building can be analyzed. In this way, monitoring allows us to integrate and increase the knowledge of the structural efficiency of an ancient building which comes from the FEM analysis.

The Vertical Displacement Monitoring of Palazzo del Capitano Pillars and Columns: The High-Precision Geometric Leveling

The vertical displacement monitoring of *Palazzo del Capitano* aimed to evaluate the presence of possible ongoing ground settling. It was performed through the high-precision geometric leveling of the building pillars and columns. This monitoring technique has already been applied to several historical buildings methodically, such as the Leaning Tower in Pisa [12] and the *Basilica di San Marco* in Venice [13].

Planning and Execution of the Leveling Network

The high-precision geometric leveling of *Palazzo del Capitano* was realized in six observation campaigns between September 2011 and October 2012. The elevation changes of the object points in time were defined referring to the altitude of "point 0" (the reference point supposed fixed), placed in the center of *Piazza Sordello* where the bench mark of the Military Geographic Institute IGM95 is located.

Monitoring was performed by a network made up with 26 object points and was structured into three closed loop lines leveling, see Figure 9.

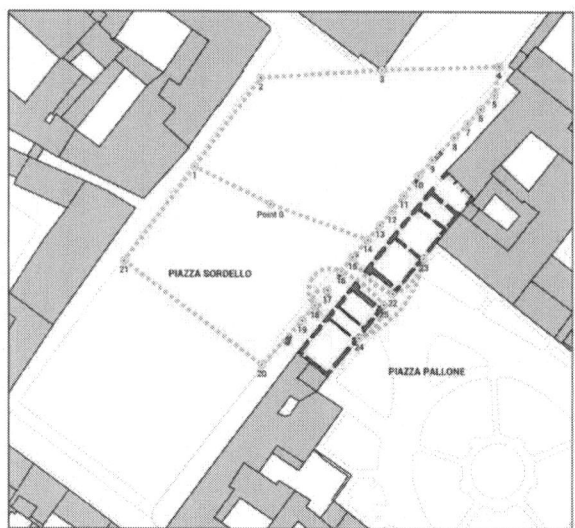

Figure 9. Plan of the leveling network with the localization of benchmarks and line leveling.

As far as monumentation is concerned, the object points were materialized with stainless steel studs (for points situated at a height lower than 1 m) and cross arms (for points situated at a height above 1 m) sealed on the structures by a bi-component easy-removal paste. In the high-precision geometric leveling it is important that all points must be well tied with the structure, otherwise recorded displacements might represent monumentation displacements instead of structure displacements.

The points were located on the exterior structures of *Palazzo del Capitano* (exterior walls and an ancient door metal hinges of the façade overlooking *Piazza Pallone*, columns basements and pillars summits of the façade overlooking *Piazza Sordello*), on some remarkable elements in *Piazza Sordello* (mainly in correspondence with some lampposts) and on the columns of the historical *Magna Domus*.

The monitoring was performed by means of rigid invar staffs with bar code reading and a Leica DNA03 digital level, which has a standard deviation height measurement per 1 km double-run equal to 0.3 mm and a range for electronic measurements from 1.8 m to 110 m [14]. The detection technique used is known as "backward-forward, forward-backward".

Outcomes of the High-Precision Geometric Leveling

The results of the leveling data—processed through the Leica GeoOffice software combined with the academic software LEV in order to compensate the observations through rigorous least squares minimization—point out the presence of real structural displacements. The error for each measurement σ_Δ is considered equal to ±0.1 − 0.2 mm and then the tolerance error T is approximately 2.5 times the σ_Δ(T = ±0.25 − 0.50 mm).

Table 1 and Figure 10 show that, on columns and pillars of the *Palazzo del Capitano* and *Magna Domus*, the data recorded a general increase of all points in summer (October 2012–June 2012) and autumn (December 2011–September 2011). Otherwise, in winter (February 2012–December 2011) and spring (June 2012–February 2012), the data recorded a general decrease of the points.

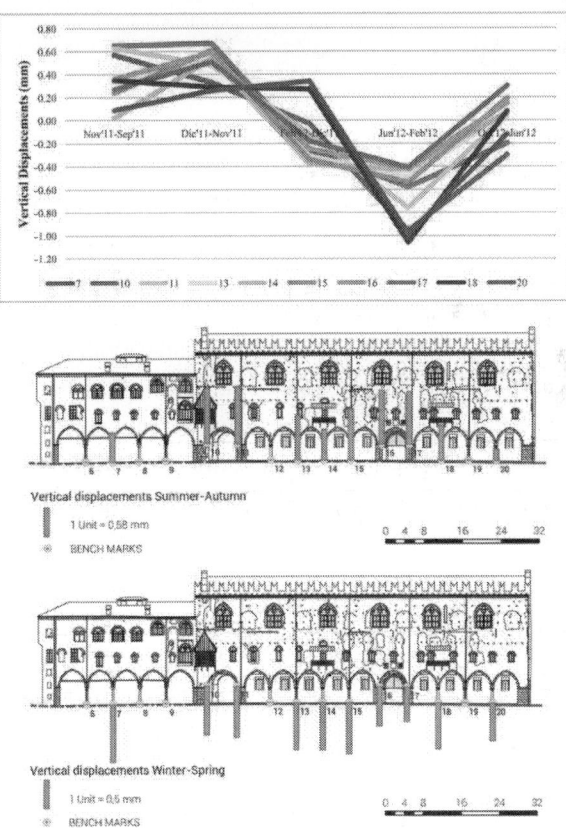

Figure 10. Evolution of the vertical displacements of the *Palazzo del Capitano* and *Magna Domus* benchmarks.

Table 1. Partial and total vertical displacements of the leveling bench marks on the columns and pillars of *Palazzo del Capitano* and adjacent *Magna Domus*. Missing data are due to benchmark removal.

Bench-Mark	Partial Displacements 11/11–09/11 (mm)	Partial Displacements 12/11–11/11 (mm)	Partial Displacements 02/12 12/11 (mm)	Partial Displacements 06/12 02/12 (mm)	Partial Displacements 10/12 06/12 (mm)	Total Displacements 10/12 09/11 (mm)
5	0.65	0.35	-0.03	---	---	---
6	0.57	0.33	-0.03	0.96	0.30	0.39
7	0.58	---	---	---	---	---
8	0.48	0.52	---	---	---	---
9	0.65	0.67	-0.26	-0.58	-0.20	0.28
10	0.63	0.48	-0.12	-0.76	0.05	0.28
11	0.54	---	---	---	---	---
12	0.33	0.52	-0.31	-0.50	0.07	0.11
13	0.02	0.54	-0.58	-0.43	0.07	0.18
14	0.24	0.61	0.34	0.54	0.19	0.16
15	0.36	0.59	-0.25	-0.42	0.15	0.43
16	0.27	0.51	0.18	-0.41	0.30	0.49
17	0.35	0.29	0.27	1.06	0.08	0.07
18	0.24	0.26	---	---	---	---
19	0.09	0.26	0.34	-0.99	-0.12	-0.42

As a result, in one year, the vertical displacements tend to decrease generally to lower themselves. Only in the correspondence of the pillars (points 9–10 and 15–16) and of the outer side of the building (points 6 and 19), there are any significant elevation changes (values bigger than the tolerance T). For this reason, the displacement field appears due to a cyclical seasonal variations (thermal phenomena) and then there aren't any sinking foundations, sub-pressures or elevation changes of the storage capacity.

Limits and Potentialities of the High-Precision Geometric Leveling Technique

The high-precision geometric leveling is an excellent monitoring tool: it allows us to detect very small vertical displacements with a high degree of accuracy, thanks to the good control that the operator can exercise on the measuring operations. In the *Palazzo del Capitano*, the reliability of the results was also guaranteed by the site conditions: it was possible to carry out an effective distribution of the object points detected under the best conditions, as it was proved by the magnitude of the closing errors of the loop lines leveling network. The only real problem for the execution of the observation campaigns was due to some vandalism exercised by passers-by who, in time, have removed some studs, causing the loss of data on some columns.

The Horizontal Displacements Monitoring of Palazzo del Capitano Main Façade: From Total Station to TLS

The horizontal displacement monitoring of *Palazzo del Capitano* was aimed at evaluating the ongoing increases in the permanent out of plane displacements of the front façade overlooking *Piazza Sordello*. The horizontal displacements monitoring of the *Palazzo del Capitano* main façade was performed according to two different closely related approaches.

First of all, a Total Station topographic monitoring to detect displacements of the spatial coordinates of 24 points, placed between the windows of the first and the second floor, was performed. This technique, ordinarily used for monitoring operations of any kind of structures (galleries, historic buildings and monuments, bulkheads, retaining walls, landslides, decks…), has subsequently paved the way for the TLS monitoring testing.

Differently from the Total Station topographic monitoring, which allows us to control only a limited number of well-identified points of the investigated surface, the monitoring with TLS could allow us to extend the control over large portions of the surface and, potentially, on the whole surface, which is clearly described by a dense point cloud acquired with the highest achievable accuracy.

The TLS technique for monitoring displacements adopts an area-based method. The deformation analysis results from a comparison between two scans acquired at different times by considering mathematical surfaces fitting measured points. In this way noise could be reduced. Given that the different point clouds have been already georeferenced to the same reference system, the comparison between interpolated surfaces allows us to evaluate displacements [15].

Introduction of TLS Monitoring

The Laser Scanning technology for the measurement of structural deformation was adopted for the first time at the beginning of the twenty-first century. Previously, in fact, the supposed lack of accuracy of the TLS system in measurement of individual points had precluded its use for monitoring applications.

Gordon et al. [16] experimented with TLS to detect the surface deformations of wooden and concrete beams subjected to loads. They confirmed the reliability of the laser scanner results compared with the measurements acquired by means of photogrammetric techniques; in this way they proved that the theoretical accuracy of the TLS on a surface generated by a dense point cloud is much greater than the precision attended from the detection of a single point. The fitting surfaces generated by the approximation of the acquired points allow us to compensate the errors on individual points, improving the general accuracy of the data.

However, till now the TLS technique for monitoring displacements has been performed only for the control of large structures and for geological applications related to the landscape. It has been used in situations where problems of particular structures of the landscape might represent a serious danger for human life and might cause an immense damage in terms of economic losses. This is the case of the large water dams [15,17], locks [18,19], or landslides [20] and quarries [21]. Therefore, the TLS monitoring of an architectural structure belonging to the historical and artistic heritage is a new field of investigation.

Actually, in the last few years, some researchers have already tested the possibility of using the potential of laser scanning for deformation monitoring of historical buildings. In 2009, Van Genechten *et al.* [22] published the results of the tests to evaluate the accuracy of TLS monitoring in detecting the deformations induced on a masonry arch. However, in this case, the experiments were performed in laboratory, under ideal conditions. Currently, the existing literature reports that the use of laser scanning for deformation monitoring of structures has actual utility for the detection of seasonal deformations of points recording displacements in the range of few centimeters [23]; this is what is expected in the case of*Palazzo del Capitano*. In other cases, a possible way of monitoring very complex architecture—integrating different instrumentation and modeling methods—is permitted by the use of TLS instrumentation [24]. The development that this process will support will be the transformation from model to 3D BIM product for structural analysis [25].

Planning and Execution of the Topographic Network and Laser Scanning Monitoring

The monitoring of the main façade of the building was realized in seven observation campaigns between October 2011 and October 2012. The topographic monitoring was performed with a Leica TCRA1203 Total Station (a particular model of the Leica TPS1200+). Data georeferencing in the ETRF2000 Reference System (RS) with UTM projection was performed by a rigid roto-translation with respect to the vertex 1000 and the direction defined by the vertices 1000–1001 (in this way the modulus of linear deformation is always equal to 1). These two points, placed in *Piazza Sordello*, have known coordinates identified by the IGM national network (point 1000 is in correspondence of the ASS2-GPS and point 1001 is in correspondence of the point ASS4-GPS). Then, from this reference system, the direction 1000–2000 was defined and it has been considered fixed in the following stage of topographic analysis for the definition of measurements with the Total Station starting from the vertices 1000, 2000, 3000 and 4000, leaving out the 1001 vertex, see Figure 11. The station vertices were implemented with steel studs sealed on the paving.

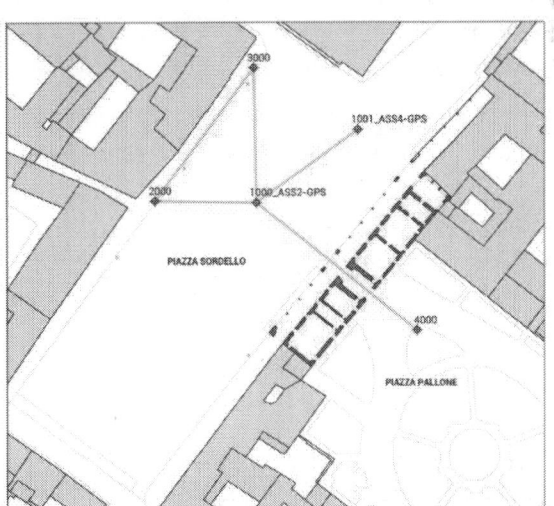

Figure 11. Plan of the topographic network with the localization of vertices.

The points of detail, measured from the vertices, were arranged on the building façade, for the displacement monitoring, and around the squares, for georeferencing system. The collimation of the network vertices has been realized by means of mini reflective prisms which have been collimated with infrared remote measure setting (IR-Mode) and the automatic target recognition.

The 24 control (CTRL) points on the building façade, were identified by plasticized reflecting square shapes (4 × 4 cm) sealed by a bi-component easy-removal paste on the sides of the windows that can be opened, see Figure 12. CTRL points were collimated with IR-Mode, without the automatic target recognition because of the distance of the points from the Total Station. In standard mode, the IR-Mode accuracy (standard deviation σ) is equal to 1 mm and the distance measurement range for mini prism is 1,200 m and for reflective tape is 250 m [14].

Figure 12. Individuation of the points of detail detected with Total Station on the *Palazzo del Capitano* façade overlooking*Piazza Sordello*.

The points of detail for cloud georeferencing (TLS points) were identified by paper targets that were placed and then removed during several observation campaigns. TLS points were collimated with reflector less remote measure setting (RL-Mode), with a Pin Pointer R400. For distances shorter than 500 m, the RL-Mode accuracy (standard deviation σ) is equal to 2 mm [14]. Processing data points software (Cyclone) recognizes TLS target vertices automatically.

Every point was detected by the Total Station twice (with both direct and reversed modes) in order to eliminate instrumental errors and increase precision.

At every observation campaign, the TLS monitoring was performed placing the laser scanner in the center of *Piazza Sordello*, in correspondence of the IGM point ASS2-GPS: in this way it was in a central position with respect to the building façade, at a distance of about 28 m from the median line of the front plane.

The compact Leica HDS6000 Laser Scanner in the first campaign and the next model Leica HDS7000in the following campaigns were used, respectively.. Both models are phase-based, dual axis compensated, ultra-high speed laser scanners, with survey-grade accuracy and field-of-view. In particular, the HDS6000 has a range up to 79 m, a scan rate up to 500,000 points/s, a modeled surface precision/noise at 25 m between 2.0–3.0 mm and at 50 m between 4.0–7.0 mm and 40,000 points/360° detected in "ultra high" scan resolution. The HDS7000 has a range up to 187 m, a scan rate up to 1,016,727 points/s, a range noise at 25 m between 0.5–1.0 mm and at 50 m between 0.8–2.7 mm and until 100,000 points/360° detected in "extremly high" scan resolution [14].

To eliminate the noises that could have been caused by the superimposition of multiple scans, a single scan, that included the entire façade of the *Palazzo del Capitano* overlooking *Piazza Sordello*, was performed. In this way the acquired TLS data were more defined. Scans were set at the highest resolution possible, available time permitting. Both HDS6000 and HDS7000 Laser Scanners were set up with ultra-high scans resolution. In particular, the HDS7000 laser acquisitions were entirely performed with a normal quality, with windows in high quality only in correspondence to the TLS points, in order to obtain an higher precision in the recognition of targets in Cyclone and, as a consequence, in the point cloud georeferencing.

In particular, the registration errors of the TLS targets were lightly superior than the standard deviation σ of the points calculated by the least squares compensation of the topographic data; however they are in the range of tolerance error T ($1\sigma = \pm2$ mm; $T_{(95\%)} = 2\sigma = \pm4$ mm; $T_{(99\%)} = 3\sigma = \pm6$ mm). As an example, the Table 2 show the registration errors obtained in the acquisition of February 2012.

Table 2. Registration errors recorded in the acquisition of February 2012.

ID	Error Vector (m)			Error (m)
	x	y	z	
1	0.002	−0.002	0.003	0.004
2	0.000	0.001	−0.006	0.006
3	0.001	−0.003	0.005	0.006
4	−0.004	0.001	0.001	0.004
5	0.001	−0.002	−0.001	0.003
6	0.002	0.003	−0.005	0.006
7	−0.002	0.001	0.003	0.004

To perform the data analysis, it was necessary to change the RS to understand the horizontal displacements of the façade: in the UTM-ETRF2000 RS the façade plane was sloping with respect to the axes East-North (X-Y), see Figure 13.

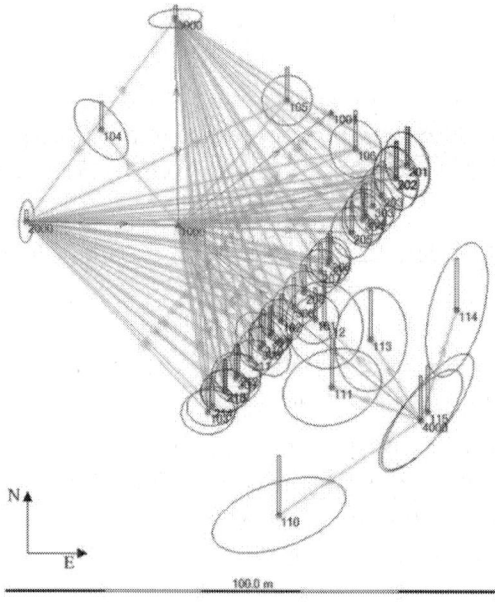

Figure 13. The October 2011 topographic network georeferenced in UTM-ETRF2000 RS. The magnification factor of the error ellipses compared with the network is equal to 5,000.

In the new defined local 3D coordinate system, all points were rigidly roto-translated to have the building façade in a parallel position to the North direction (Y axis) and the point 1000 in the position (0,0,0). So the horizontal displacements were easily readable along the East direction (X axis), see Figure 14 and Table 3 and Table 4. Then, for processing data by means of 3D Analyst Tools in ArcGIS software, a rotation of the Cartesian axes was performed to have the building façade in the XY plane and the out of plane displacements readable along the Z axis (with direction from *Piazza Pallone* to *Piazza Sordello*), see Figure 15.

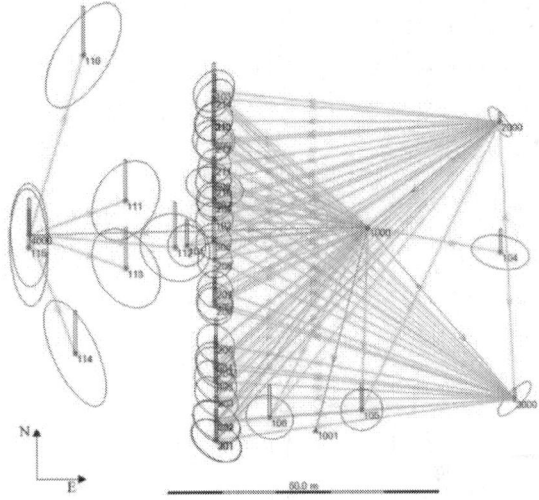

Figure 14. The October 2011 topographic network after the roto-translation of all points in a new 3D coordinate system. The magnification factor of the error ellipses compared with the network is equal to 5,000.

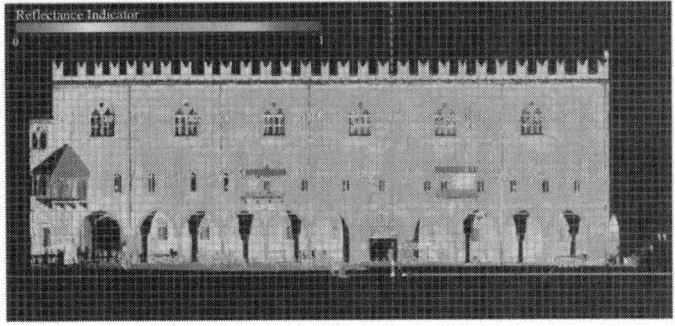

Figure 15. Cyclone. Point cloud of the *Palazzo del Capitano* façade in the new local reference system with the axes rotated. Hue expresses the reflectance of the surface materials acquired by the TLS.

Table 3. Coordinates and standard deviations of the points of detail detected on the *Palazzo del Capitano* façade. Data refer to October 2011 in new 3D coordinate system.

Bench Mark	Coordinates (m)			Standard Deviations (m)		
	E	N	Elevation	E	N	Elevation
201	971.8506	968.6920	33.3835	0.000592	0.000584	0.000528
202	971.8702	961.1654	32.6732	0.000579	0.000573	0.000508
203	971.8640	969.1189	32.6706	0.000566	0.000558	0.000487
204	971.8786	974.4860	32.7069	0.000554	0.000536	0.000463
205	971.8709	978.7395	32.6572	0.000552	0.000515	0.000446
206	971.7867	985.7351	33.9247	0.000557	0.000473	0.000430
207	971.8088	988.7880	32.6496	0.000566	0.000454	0.000414
208	971.7806	994.0845	32.6549	0.000578	0.000425	0.000405
209	971.7555	1,004.4292	32.5060	0.000583	0.000415	0.000403
210	971.7532	1,007.3552	33.8211	0.000966	0.000540	0.000563
211	971.7390	1,011.2715	32.4856	0.000573	0.000449	0.000415
212	971.7583	1,015.3628	32.4675	0.000566	0.000475	0.000427
213	971.8111	1,019.4217	32.4717	0.000563	0.000500	0.000441
214	971.8500	1,023.5099	32.4798	0.000564	0.000523	0.000458
301	971.7989	961.1570	40.1443	0.000590	0.000583	0.000548
302	971.8681	964.5062	40.2765	0.000577	0.000571	0.000533
303	971.8305	971.9479	40.1427	0.000555	0.000543	0.000504
304	971.8007	975.2991	40.1139	0.000548	0.000528	0.000493
305	971.6300	986.4897	40.1335	0.000547	0.000467	0.000469
306	971.6133	997.5203	40.1605	0.000562	0.000414	0.000461
307	971.6313	1,005.1214	40.0876	0.000562	0.000418	0.000461
308	971.6043	1,008.4506	39.9984	0.000559	0.000432	0.000464
309	971.6762	1,016.0327	40.0491	0.000646	0.000525	0.000529
310	971.7189	1,019.3767	40.0166	0.000553	0.000496	0.000483

Table 4. Error ellipses of the points of detail detected on the *Palazzo del Capitano* façade. Data refer to October 2011 in new 3D coordinate system.

Bench Mark	Error Ellipses (m)			
	Semi-Major Axis	Semi-Minor Axis	Azimuth of Major Axis	Elevation
201	0.001456	0.001422	68.82	0.001034
202	0.001445	0.001373	56.64	0.000996
203	0.001436	0.001311	54.84	0.000955
204	0.001429	0.001233	57.47	0.000908
205	0.001427	0.001173	61.52	0.000874
206	0.001422	0.001087	71.08	0.000842
207	0.001429	0.001055	76.15	0.000812
208	0.001431	0.001016	86.10	0.000794
209	0.001433	0.001010	107.19	0.000790
210	0.002422	0.001210	116.22	0.001104
211	0.001432	0.001059	119.83	0.000814
212	0.001434	0.001103	126.21	0.000836
213	0.001437	0.001153	131.51	0.000864
214	0.001443	0.001210	135.70	0.000898
301	0.001450	0.001421	72.60	0.001074
302	0.001435	0.001375	58.06	0.001045
303	0.001414	0.001269	56.18	0.000989
304	0.001406	0.001222	58.44	0.000967
305	0.001389	0.001082	72.46	0.000919
306	0.001378	0.001007	92.97	0.000903
307	0.001380	0.001017	108.28	0.000905
308	0.001384	0.001038	114.51	0.000909
309	0.001662	0.001178	128.76	0.001037
310	0.001405	0.001154	130.91	0.000946

Data Analysis of the Total Station Monitoring

From the Total Station monitoring of the CTRL points on the *Palazzo del Capitano* façade, the component of the displacement vector in the out of plane direction (Z axis) is analized.

In particular,:

- partial displacements are in the range of a few millimeters;
- total displacements generally tend to be withdrawn in the limits of the instrumental precision ($\sigma = \pm1$ mm on average) evaluated with a tolerance T $= \pm2$ mm. This would mean a general recovery of displacements detected in the partial measurements.

The outcomes of the topographic monitoring with Total Station in z direction are recorded in Table 5.

Table 5. Partial and total displacements in the out of plane direction of the points of detail detected on the *Palazzo del Capitano* façade.

Bench Mark	Partial Displacement 11/11–10/11 (mm)	Partial Displacement 12/11–11/11 (mm)	Partial Displacement 02/12–12/11 (mm)	Partial Displacement 05/12–02/12 (mm)	Partial Displacement 06/12–05/12 (mm)	Partial Displacement 10/12–06/12 (mm)	Total Displacement 10/12–10/11 (mm)
201	3.60	−2.81	1.19	−1.28	7.03	−5.39	2.34
202	2.49	−2.19	0.75	−1.48	6.85	−5.33	1.09
203	2.22	−1.99	0.48	−0.22	5.48	−4.07	1.90
204	1.58	−1.51	0.90	−1.17	5.81	−4.67	0.94
205	1.77	−2.74	0.81	0.31	4.09	−2.83	1.41
206	1.23	−0.93	−0.07	0.48	---	---	1.00
207	1.38	−1.80	−0.05	1.05	3.48	−3.14	0.92
208	1.44	−1.46	0.02	0.91	1.84	−1.93	0.82
209	−0.36	−0.88	−0.77	2.33	1.26	−1.85	−0.27
210	---	---	−0.34	2.32	1.07	−1.98	0.48
211	−0.45	−0.54	−0.63	2.31	0.56	−1.86	−0.61
212	−1.13	0.47	−0.70	2.49	0.05	−0.82	0.36
213	−0.17	0.12	−0.69	3.42	−1.05	−1.18	0.45
214	−0.92	−0.30	−1.24	4.24	−1.79	0.62	0.61
301	1.60	−2.32	1.25	−1.37	7.24	−5.22	1.18
302	1.53	−1.00	0.46	−1.44	7.39	−5.41	1.53
303	2.61	−2.24	0.92	−0.22	6.16	−4.14	3.09
304	1.82	−0.12	−0.47	−0.17	5.62	−4.64	2.04
305	0.15	−0.74	0.31	−0.22	3.69	−3.27	−0.08
306	0.27	−0.95	−0.03	1.64	0.40	−2.40	−1.07
307	−0.19	−0.48	−0.06	1.80	−0.69	−1.74	−1.36
308	0.26	−1.10	−0.53	2.90	−1.21	−1.91	−1.59
309	−0.38	−2.37	1.28	3.08	−1.66	−1.23	−1.28
310	−0.58	−0.79	−1.05	3.78	−1.27	−0.67	−0.58

Data Analysis of the Laser Scanning Monitoring

According to the area-based method, mathematical surfaces fitting measured points on the basis of the Inverse Distance Weighted (IDW) algorithm (raster grid with a constant mesh size) —through subtraction algebraic functions—were generated. Then, the displacements occurred in time were calculated in ArcGIS from two georeferenced scans acquired at two different times.

For data management reasons, the analysis was only performed on two portions of the masonry: two slices at two different heights of the façade with a thickness of 1 m and characterized by homogeneous masonry without discontinuities. The first one was placed above the windows of the second floor, the second one was placed between the first and second floors.

From the data analysis of the two slices under consideration, it is possible to observe:

- partial displacements are generally bigger than one centimeter (in the slices placed between the first and second floors, the partial displacements are in the range of +1.4/−2.1 cm);
- total displacements are calculated in the range of +7/−9 mm: this would mean the presence of a residual deformation of the masonry at the end of the entire observation period.

Figure 16 and Figure 17 show the data elaboration of the lower masonry slice analyzed, in relation to the total displacements.

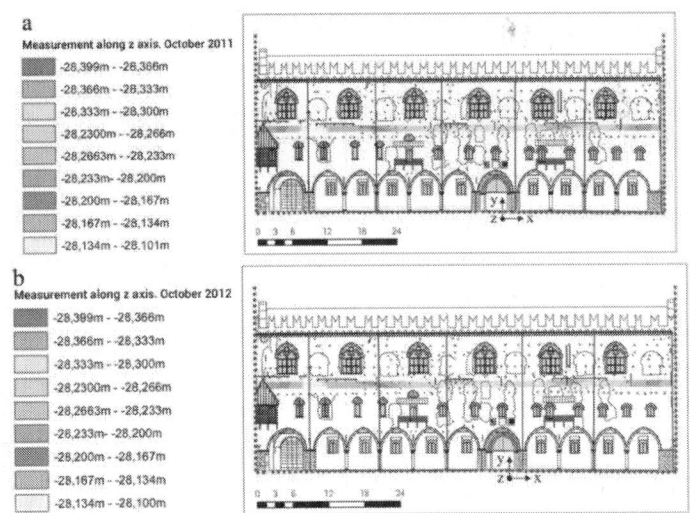

Figure 16. TLS data analysis in ArcGIS software: method of points interpolation based on the Inverse Distance Weighted (IDW) algorithm (**a**) October 2011. (**b**) October 2012.

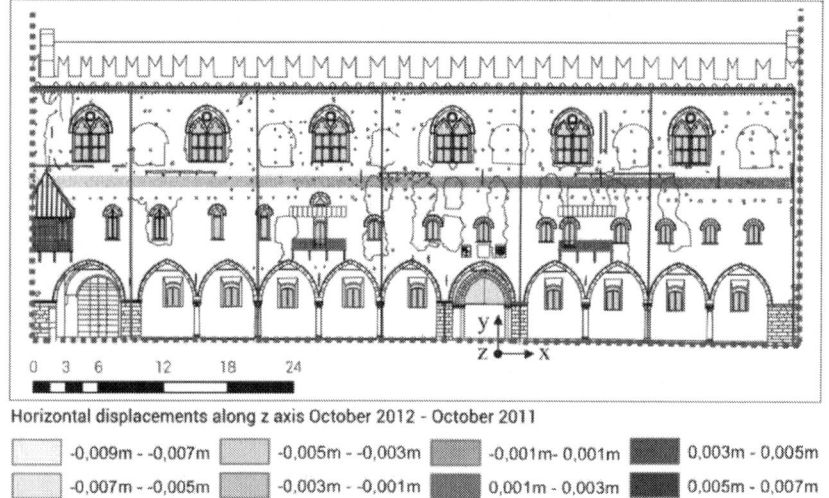

Horizontal displacements along z axis October 2012 - October 2011

-0,009m - -0,007m	-0,005m - -0,003m	-0,001m- 0,001m	0,003m - 0,005m
-0,007m - -0,005m	-0,003m - -0,001m	0,001m - 0,003m	0,005m - 0,007m

Figure 17. TLS data analysis in ArcGIS software: horizontal displacements calculation by means of the area-based method.

Synchronic Reading of the Results

From the comparative analysis of the monitoring results with Total Station and TLS, a still not exhaustive picture of the situation is evident. However, some reflections on results and procedures adopted can induce us to look for some data interpretations that may contribute to the future development and improvement of the tested monitoring techniques.

Generally, it is possible to observe that the deformation data detected with the two monitoring methods tend to diverge significantly. The Total Station monitoring records horizontal displacements in the range of a few millimeters; otherwise, the laser scanning monitoring records horizontal displacements in the range of few centimeters.

The main factor that could have affected the results is probably due to the possible inaccuracies in CTRL points acquisition with the Total Station. In particular:

- possible laying errors committed by the operator. As mentioned above, because of the distance between the network vertices and the targets on the building façade, the measurements were performed without automatic target recognition: this didn't allow the operator to eliminate possible laying errors. In view of the magnitude of the displacements, even a small imprecision in centering the targets could have caused a significant error in data detection. However, now, there is no possibility to suppress this uncertainty;

- stability of the reference system. The existing literature on the use of laser scanning technology for structural monitoring states that the reliability of the deformations estimate is closely related to the accuracy of georeferencing operations. Certainly, in the case study under consideration, the materialization of a stable reference system was the first problem to influence the results accuracy. As above mentioned, in the *Palazzo del Capitano* case, the vertices of the topographic network were implemented with steel studs sealed on the paving, which were centered through the laser plumb of the total station. Under ideal conditions, it would be necessary to materialize the vertices through the provision of reinforced concrete plinths, on which rigidly fixing metal references to the precision keying of the metal pillars where running the forced centering with total station. These arrangements would guarantee the stability of the reference system. For example, the method described above, was performed for the deformations monitoring of the Cancano dam [15], of the landslides of two rock faces in the lombard Prealps [20], and for the control of the inclination of the Leaning Tower of Pisa. Otherwise, in the case study of the *Palazzo del Capitano*, the monumentation of the reference system wasn't possible due to the constraints imposed by the local Superintendence for the Architectural Heritage, for the monumental complex of *Piazza Sordello*.

CONCLUSIONS

In this paper, some significant considerations about the combined use of surveying and monitoring as tools to increase the knowledge about an historical building were presented.

Firstly, it has been underlined the importance of having a geometrical survey that can represent in the most complete, accurate and truthful way an analyzed situation. The information provided by the geometrical survey are the basis for the definition of a reliable structural 3D model, adopted to simulate the expected behavior of the building in normal operating conditions and in respect of exceptional events according to the current codes.

In this contest, the potentialities offered by the use of a TLS to the survey of an historical building were evaluated. It was observed how this technology allows us to minutely and completely restore all the irregularities and anomalies of the acquired objects, enabling the construction of very realistic 3D models. Otherwise, problems related to the use of TLS data for the structural analysis were analyzed, too. These problems are mainly due to the way the metrical information can be transferred in a simplified 3D model, which is able to simulate realistically and in a useful and acceptable time the building structural behavior.

Secondly, the attention has been focused on the importance of monitoring to understand the real structural behavior of the building (and then recording the real displacements) in a medium-long term period. In this context different monitoring techniques have been integrated. Some of them have already been tested in time—high-precision geometric leveling to measure vertical displacements and topographic monitoring to measure horizontal displacements of some well-identified points in the main façade—others have yet to be tested—TLS monitoring to measure horizontal displacements of entire portions of the main façade. As it was often underlined, the TLS monitoring technique has already been applied for the control of movements in large structures and in particular landscape situations; however its application in civil structures has to be verified yet.

The experience with the *Palazzo del Capitano* is a starting point to test the efficiency of the laser scanner technique as a monitoring tool for historical buildings. Given that, on the territory investigated, there is a considerable number of building belonging to the national and international cultural heritage and given that significant seismic events occurred last year in the Mantua area, in the near future there will be the possibility to test thoroughly and systematically the real efficiency of the laser scanner technique for the control and the protection of civil structures of significant historical and cultural relevance. From the outcomes and the experience reported in the current paper, it is possible to attest to the methodology and rigor needed to encourage improvements in the TLS monitoring technique, in order to achieve more reliable results in the detection of the structural movements of the investigated buildings, in the future.

The authors express their thanks to L'Occaso Stefano—Director of *Palazzo Ducale* in Mantova—and to Paolozzi Strozzi Giovanna—*Soprintendente BSAE* in Mantua, Brescia and Cremona—for having contributed to this work by making available all the information and data at their disposal.

CONFLICT OF INTEREST

The authors declare no conflict of interest.

REFERENCES

1. *Nuove Norme Tecniche per le Costruzioni, New Technical Standards for Construction*; DM Ministerial Decree: Roma, Italy; 14; January; 2008.
2. *Assessment and Reduction of the Seismic Risk of Cultural Heritage*; Technical Standards for Construction; Ministry of Intrastructure: Roma, Italy; 14; January; 2008.
3. Castagnetti, C.; Bertacchini, E.; Capra, A.; Dubbini, M. Terrestrial Laser Scanning for Preserving Cultural Heritage: Analysis of Geometric Anomalies for Ancient Structures. Proceedings of the FIG Working Week 2012 on Knowing to Manage the Territory, Protect the Environment, Evaluate the Cultural Heritage, Rome, Italy, 6–10 May 2012.
4. Pesci, A.; Bonali, E.; Galli, C.; Boschi, E. Laser scanning and digital imaging for the investigation of an ancient building: Palazzo d'Accursio study case (Bologna, Italy). *J. Cult. Herit* **2012**, *13*, 215–220.
5. Pesci, A.; Casula, G.; Boschi, E. Laser scanning the Garisenda and Asinelli Towers in Bologna (Italy): Detailed deformation patterns of two ancient leaning buildings. J. Cult. Herit. **2011**, *12*, 117–127.
6. Achilli, V.; Bragagnolo, D.; Fabris, M.; Menin, A.; Salemi, G. Metodologie Geomatiche per il Rilievo Integrato Finalizzato alla Modellazione Strutturale. Proceedings of the 9th National Conference ASITA, Catania, Italy, 15–18 November 2005; pp. 15–24.
7. Cardaci, A.; Mirabella, R.G.; Versaci, A. From the Continuos to the Discrete Model: A Laser Scanning Application to Conservation Projects. Proceedings of the International Archives of the Photogrammetry, Remote

Sensing and Spatial Information Sciences, Trento, Italy, 2–4 March 2011; Volume 38, pp. 437–444.

8. Camarda, M.; Guarnieri, A.; Milan, N.; Vettore, A. Health Monitoring of Complex Structure Using TLS and Photogrammetry. Proceedings of the International Archives of the Photogrammetry, Remote Sensing and Spatial Information Sciences, Newcastle upon Tyne, UK, 21–24 June 2010; Volume 38, pp. 125–130.

9. Fabris, M.; Achilli, V.; Bragagnolo, D.; Menin, A.; Salemi, G. Laser Scanning Methodology for the Structural Modelling. Proceedings of the XXI International International Committee for Architectural Photogrammetry Symposium (CIPA), Athens, Greece, 1–6 October 2007.

10. *ABAQUS/Standard, Theory and Users Manuals, Release 6.10-1*; HKS Inc.: Pawtucket, RI, USA, 2007.

11. Barbieri, G.; Biolzi, L.; Bocciarelli, M.; Fregonese, L.; Frigeri, A. Assessing the seismic vulnerability of an historical building. *Eng. Struct.* **2013**. in press.

12. Burland, J.B.; Viggiani, C. Osservazioni sul comportamento della Torre di Pisa. *Riv. Ital. Geotec.***1994**, *38*, 179–200.

13. Giussani, A.; Monti, C. Basilica di San Marco: Operazioni di misura per il controllo statico. *Riv. Catasto dei Serv. Tec. Erar.* **1985**, *1*, 3–20. [Google Scholar]

14. Leica Geosystem. Available online: http://www.leica-geosystems.it/it/index.htm (accessed on 21 May 2013).

15. Alba, M.; Fregonese, L.; Prandi, F.; Scaioni, M.; Valgoi, P. Structural Monitoring of a Large Dam by Terrestrial Laser Scanning. Proceedings of the ISPRS Commission V Symposium, Dresden, Germany, 25–27 September 2006.

16. Gordon, S.J.; Lichti, D.; Stewart, M.; Franke, J. Structural Deformation Measurement Using Terrestrial Laser Scanners. Proceedings of the 11th International FIG Symposium on Deformation Measurements, Santorini Island, Greece, 25–28 May 2003.

17. González–Aguilera, D.; Gómez–Lahoz, J.; Sánchez, J. A New approach for structural monitoring of large dams with a three-dimensional laser scanner. *Sensors* **2008**, *8*, 5866–5883.

18. Schäfer, T.; Weber, T.; Kyrinovič, P.; Zámečniková, M. Deformation Measurement using Terrestrial Laser Scanning at the hydropower station of Gabcikovo. Proceedings of the 3rd International Conference on Engineering Surveying and FIG Regional Conference for Central and Eastern Europe (INGEO 2004), Bratislava, Slovakia, 11–13 November 2004.

19. Lindenbergh, R.; Pfeifer, N. A Statistical Deformation Analysis of Two Ephocs of Terrestrial Laser Data of a Lock. Proceedings of the 7th

Conference on Optical 3-D Measurement Techniues, Vienna, Austria, 3–5 October 2005; pp. 61–70.

20. Alba, M.; Roncoroni, F.; Scaioni, M. Application of TLS for Change Detection on Rock Faces. Proceedings of the International Archives of the Photogrammetry, Remote Sensing and Spatial Information Sciences, Paris, France, 1–2 September 2009; Volume 38, pp. 99–104.

21. Scaioni, M.; Giussani, A.; Roncoroni, F.; Sgrenzaroli, M.; Vassena, G. Monitoring of a Geological Sites by Laser Scanning Techniques. Proceedings of the International Archives of the Photogrammetry, Remote Sensing and Spatial Information Sciences, Istanbul, Turkey, 12–23 July 2004; Volume 35, pp. 708–713.

22. Van Genechten, B.; Demeyere, T.; Herinckx, S.; Goos, J.; Schueremans, L.; Roose, D.; Santana, M. Terrestrial Laser Scanning in Architectural Heritage—Deformation Analysis and the Automatic Generation of 2D Cross-Sections. Proceedings of XXII International International Committee for Architectural Photogrammetry Symposium on Digital Documentation, Interpretation & Presentation of Cultural Heritage (CIPA), Kyoto, Japan, 11–15 October 2009.

23. Alba, M.; Barazzetti, L.; Giussani, A.; Roncoroni, F.; Scaioni, M. Sperimentazione di tecniche innovative per il monitoraggio delle strutture. Proceedings of Geomatics, the roots of the future. Tribute to Sergio Dequal and Riccardo Galetto, Pavia, Italy, 10–11 February 2011.

24. Fassi, F.; Achille, C.; Fregonese, L. Surveying and modelling the main spire of Milan Cathedral using multiple data sources. *Photogramm. Rec.* **2011**, *26*, 462–487.

25. Achille, C.; Fassi, F.; Fregonese, L. 4 Years History: From 2D to BIM for CH—The Main Spire on Milan Cathedral. Proceedings of the 2012 18th International Conference on Virtual Systems and Multimedia (VSMM 2012), Milan, Italy, 2–5 September 2012; pp. 377–382.

CITATION

Luigi Fregonese, Gaia Barbieri , Luigi Biolzi , Massimiliano Bocciarelli , Aronne Frigeri and Laura Taffurelli, Surveying and Monitoring for Vulnerability Assessment of an Ancient Building, doi:10.3390/s130809747

CHAPTER 7

A Methodology to Validate the InSAR Derived Displacement Field of the September 7th, 1999 Athens Earthquake Using Terrestrial Surveying. Improvement of the Assessed Deformation Field by Interferometric Stacking

Ioannis Kotsis [1], Charalabos Kontoes [2], Dimitrios Paradissis [1], Spyros Karamitsos [1], Panagiotis Elias [2] and Ioannis Papoutsis [2]

[1]Higher Geodesy Laboratory and Dionyssos Satellite Observatory, National Technical University of Athens, Iroon Polytexneiou 9, 15780, Zografou, Greece;
[2]National Observatory of Athens, Institute of Space Applications and Remote Sensing, Vas. Pavlou and Metaxa str., 15236, Palaia Penteli, Greece;

ABSTRACT

The primary objective of this paper is the evaluation of the InSAR derived displacement field caused by the 07/09/1999 Athens earthquake, using as reference an external data source provided by terrestrial surveying along the Mornos river open aqueduct. To accomplish this, a processing chain to render comparable the leveling measurements and the interferometric derived measurements has been developed. The distinct steps proposed include a solution for reducing the orbital and atmospheric interferometric fringes and an innovative method to compute the actual InSAR estimated vertical ground subsidence, for direct comparison with the leveling data. Results indicate that the modeled deformation derived from a series of stacked interferograms, falls entirely within the confidence interval assessed for the terrestrial surveying data.

INTRODUCTION

One of the most significant natural disasters to strike Greece in the 20th century was the September 7, 1999, $11^h 56^m 50^s$ UTC, Mw (moment magnitude) = 5.9 Athens earthquake. It claimed the lives of 143 people, and caused the collapse of several buildings, mainly in the northwest suburbs of the Greek capital. The approximate location of the earthquake epicenter was 38.10°N, 23.56°E, roughly 20 km northwest from the center of Athens [1].

The vertical displacement field at the surface level caused by this tectonic event was investigated with space born Synthetic Aperture Radar Interferometry (InSAR), using ERS-2 data. InSAR processing showed a significant deformation with the maximum Line Of Sight (LOS) subsidence being of approximately 6 cm [1]. This observation was used in earthquake modeling and fault location mapping [2-9] along the middle of the Parnitha mountain. However, the deformation field reported in [1] could not be verified at that time due to the lack of co-seismic geodetic measurements of adequate precision. The sole indication was provided by geologists and engineers who visited the area and confirmed that the damaged structures, at the substructure level, were showing a vertical movement of the same order of magnitude as the InSAR derived assessments.

The region of maximum deformation coincided with the main shock epicenter. This area was very close to the Mornos river open aqueduct, used for water supply to Athens. The distance of the aqueduct pass from the earthquake epicenter was less than 2.5 km. The water supply authority in Athens awarded an aqueduct-leveling project to the National Technical University of Athens/Higher Geodesy department (NTUA/HG), which lasted for two months, from March to April 2001. Prior leveling data along the Mornos aqueduct had been obtained in 1984. No height data were available for the intermediate time interval 1984-2001; however no major seismic event had taken place in that period. The two co-seismic sets of leveling data were considered adequate to investigate the vertical displacement in the affected by the earthquake area and verify the InSAR derived observations. Figure 1(a) illustrates the leveling path legs and the Mornos aqueduct projected onto the 1:50,000-scale map. Figure 1(b) shows the area where leveling data were acquired, projected onto the calculated interferogram. The test area extends from 38°09'N 23°31'E to 38°06'N 23°38'E.

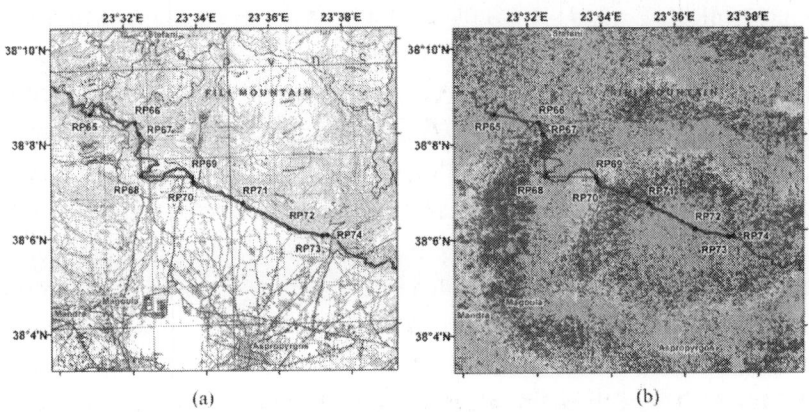

Figure 1. (a) Plots of the Mornos aqueduct (blue) and height network (red) projected on 1:50000-scale map and (b) onto an ERS-2 SAR image interferogram.

The scope of this paper is the evaluation of the InSAR derived displacement field caused by the Sept 7, 1999 Athens earthquake, using as reference an external data source provided by terrestrial surveying along the Mornos river open aqueduct. Research works relating to InSAR – leveling interoperability issues have been published in the past, focusing on either verifying the InSAR derived subsidence, or integrating them with the leveling data to increase the reliability of the measurement. In [10] a spatially dense network of leveling benchmarks was used, to integrate terrestrial measurements with InSAR data, and sums of Gaussian surfaces were proposed to approximate the subsidence field induced from oil/gas extraction activity. Moreover, in [11] a method to improve the InSAR derived deformation field was presented, by splitting the differences between InSAR and leveling derived assessments to two components: one mathematical model accounted for the mean tropospheric effects and orbital errors, and a second model was used to describe for the local, less correlated error sources, such as Digital Elevation Model (DEM) errors and local atmospheric effects. By approximating models with polynomials and by generating a non – mathematical model for the residuals of the approximations, corrections for the InSAR derived deformations were produced for the entire SAR image. In [12] a study for mine subsidence monitoring using ERS-1/2 and JERS-1/2 was investigated, combining the resulted subsidence with ground-collected data. In [13] InSAR derived deformations were compared and correlated with temporally dense leveling data for settlements monitoring in the reclaimed land of the new Hong Kong international airport and the Fairview Park.

This paper is structured as follows: section 2 refers to the preliminary processing of the input data, namely the InSAR and leveling measurements. Section 3 presents in an analytic way the distinct steps in rendering the two data sets compatible. Section 4 outlines the results obtained by applying the proposed processing chain, whilst section 5 investigates more thoroughly the physical meaning of these results and the applicability of the method in verifying InSAR derived subsidence on the basis of terrestrial surveying data.

INPUT DATA

ERS-1/2 InSAR Data

ERS-1/2 sensor images spanning the period from December 1997 to January 2001 were acquired and processed over the Athens Greater Area. The satellite images were provided by the European Space Agency in the frame of the ESA-GREECE AO project 1489OD/11-2003/72.

Interferometric calculations were done by using the CNES DIAPASON InSAR processing software, and the sixteen most coherent co-seismic interferograms were kept for the purposes of the study. The image pairs used along with their corresponding "altitude of ambiguities" are shown in Figure 2. The influence of the terrain relief on the interferograms was lifted out using a DEM, which was originated by digitizing the 20 m contour lines from the 1:50,000-scale topographic maps. The high frequency DEM artifacts remaining in the interferograms, were calculated as the ratio of the DEM error (~10 m) over the interferometric "altitude of ambiguity" (20 m–417 m) [14]. They were all estimated to be below the cycle level (0.3–0.02 cycles).

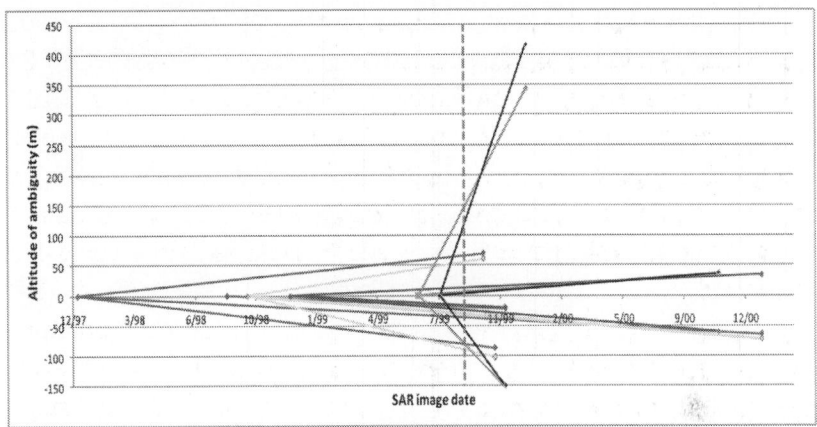

Figure 2. Set of interferometric pairs used in the study. The vertical dashed line indicates the date of the earthquake occurrence.

Leveling Data Along the Mornos Aqueduct

The first terrestrial surveying work on the aqueduct was done in 1984, covering its whole length of approximately 200 km. A special trigonometric height technique was used, providing the same level of accuracy as conventional leveling but being significantly faster [15]. This technique employed a highly accurate geodetic total station to obtain the slope distance and the vertical angle between the two points of interest. The use of a redundant number of stationary sets of tripods and tribranch adapters eliminated the need for target and instrument height measurements. Furthermore, atmospheric refraction effects were further eliminated by concurrent measurements at both ends of an observation line - leading to high accuracy observations.

Moreover, a standard geometric leveling was realized in 2001. The total distance surveyed was 40 km, of which 12 km were confined in the area of interest illustrated in Figure 1(b). Figure 3 shows the leveling path legs and the longitudinal axis of the open aqueduct, projected onto a wrapped interferogram.

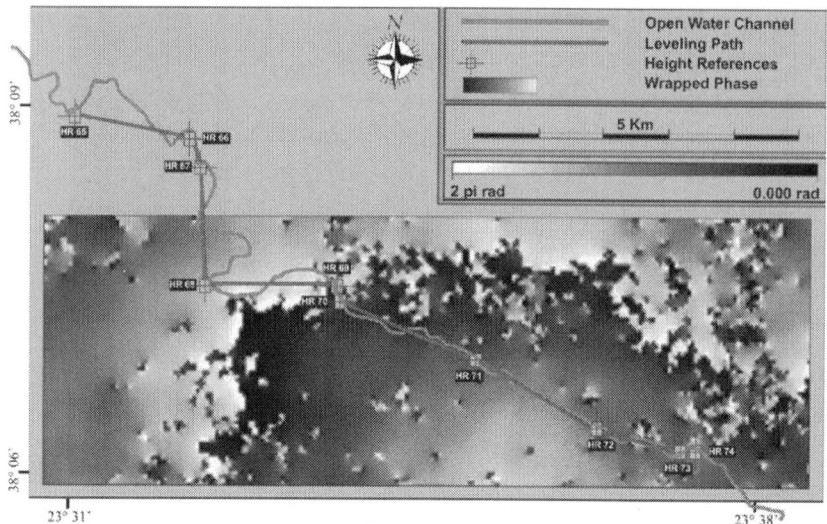

Figure 3. Leveling path legs plot (red) and aqueduct plot (blue) projected onto a wrapped interferogram. For clarity purposes, only the segments connecting the height references are displayed. The actual leveling path follows the channel.

The accuracy of the leveling works was estimated to be of the order of a few millimeters between successive height references [16]. It should be noted that the two leveling experiments conducted in years 1984 and 2001 used exactly the same height reference points. The height differences obtained by surveying the aqueduct at the two epochs indicated a significant vertical displacement induced by the earthquake. Taking into account the standard deviations of the geometric leveling and the trigonometric leveling and by applying the error propagation law, the standard deviations of the height differences were estimated to range from 4 mm to 8 mm. These values correspond to the relative heights between successive height benchmarks, depending on the length of the leveling path segments.

RENDERING INSAR DATA COMPARABLE TO LEVELING DATA

The differential displacement data derived by the two different techniques were incompatible and consequently a direct comparison was not possible. These incompatibilities may be summarized as follows:

- InSAR processing provided wrapped interferograms, consequently only the fractional part $\mod_{2\pi}\Phi$ of the full phase difference Φ was known.
- InSAR results correspond to the projection of the true vertical deformation along the LOS vector.
- The reference systems of the leveling data and the InSAR data were different. InSAR data were referring to ED 50 UTM zone 34 while the leveling data were referring to the mean sea level and the height reference positions to the Hellenic Geodetic Reference System 87 (HGRS 87).
- The interferograms were "noisy" mainly due to temporal decorrelation, orbital and tropospheric disturbances.

The following sections describe the procedure used to eliminate the effects of the above types of incompatibility, rendering the two datasets comparable.

Wrapped Interferogram Filtering

The wrapped interferogram underwent a simple filtering procedure. The primary objective of this action was to minimize the probability of phase unwrapping failure, while a secondary goal was the improvement of the wrapped and unwrapped interferogram appearance in order to derive qualitative evaluations more efficiently. The filter used was a simple 2D 3×3 space mean filter (symmetric to match the rectangular pixel dimensions), applied on both the real $\cos(\psi_{i,j})$ and imaginary $\sin(\psi_{i,j})$ parts of a virtual unitary magnitude signal $e^{j\psi_{i,j}} = \cos(\psi_{i,j}) + j\sin(\psi_{i,j})$. The phase of this signal is the unfiltered interferometric phase $\psi_{i,j}$ In other words, the 2D space filter was applied on a unitary signal to which the phase of the input interferogram was projected. The phase $\psi_{flt/i,j}$, comprising the filtered interferogram, was extracted through an arctan operation from the filtered real and imaginary parts of the virtual signal. The filtering procedure is best defined by the following formula:

$$\Psi_{flt/i,j} = \arctan\left(\sum_{j=j_0-\frac{k-1}{2}}^{j=j_0+\frac{k-1}{2}}\sum_{i=i_0-\frac{k-1}{2}}^{i=i_0+\frac{k-1}{2}}\frac{\sin(\psi_{i,j})}{k^2}\Bigg/\sum_{j=j_0-\frac{k-1}{2}}^{j=j_0+\frac{k-1}{2}}\sum_{i=i_0-\frac{k-1}{2}}^{i=i_0+\frac{k-1}{2}}\frac{\cos(\psi_{i,j})}{k^2}\right)$$

(1)

where k is the filter size, which equals 3. This value was considered to be an optimal one, as it corresponds to a satisfactory tradeoff between interferometric spatial resolution and level of smoothing. The criterion for choosing k was to eliminate isolated pixel noise while keeping the spatial deformation trend evident in the interferogram. In Figure 4, the effect of the interferogram filtering procedure is presented.

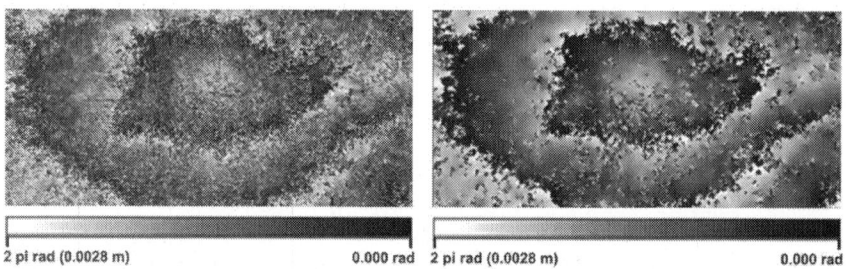

2 pi rad (0.0028 m)   0.000 rad   2 pi rad (0.0028 m)   0.000 rad

Figure 4. Wrapped interferogram, before (left) and after (right) filtering.

Phase Unwrapping

Various 2D phase unwrapping techniques have been developed for resolving the "integer ambiguity" problem of the interferometric phases. In this study "Quality Guided Path Following", "Least Squares Without Weights", "Weighted Least Squares", and "Minimum LP Norm" approaches were implemented [17-20]. The unwrapped interferograms produced by these techniques were evaluated for surface discontinuities, by inspecting for the presence of breaklines (abrupt gradient changes) or "tears" (non – derivabilities) and measuring their length. As a result, it was inferred that the most effective technique, for this particular scenario, was the "Weighted Least Squares". The weights were derived from the coherence map, representing the computed cross correlation between the master and the slave image.

The unwrapped co-seismic interferograms were all undergone a special processing in order to minimize the existing orbital, tropospheric and DEM disturbances. These errors were lifted by a "tilting" and "shifting" operation, using a number of coherent pixels located outside the deformed area. According to this approach [21], the deformation on these pixels was expected to follow a well-defined t-student distribution around a local zero mean. Then, by forcing each local deformation mean to zero, the calculated interferograms were "tilted" and "sifted". Figure 5 emphasizes the effect of this process, where the disposal of the orbital fringes becomes evident.

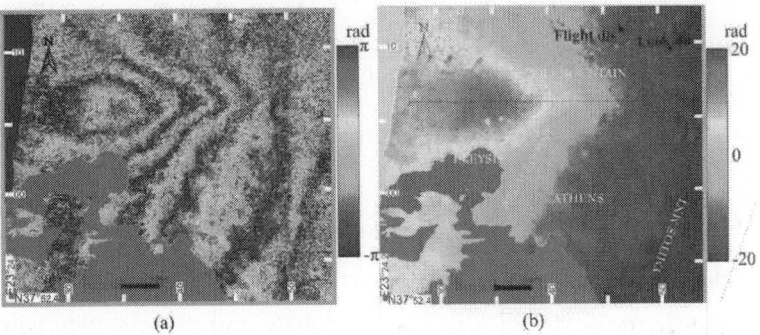

(a)   (b)

Figure 5. (a) & (b) Wrapped and unwrapped versions of the same interferogram. Note the unrealistic fringe pattern due to inaccuracies in the orbital data used. (c) & (d) The effect of the "tilting" and "shifting" operation on the same interferogram; the orbital fringes are removed.

Incidence Angle Correction

The HGRS 87 unwrapped interferogram provides the differential vertical displacements for each target pixel as projected to the LOS vector $\Phi_{LOS}(E,N)$, and not the vertical differential displacements $\Phi_{dU}(E,N)$ themselves, as is the case of leveling (Figure 6). These two quantities are related through the incidence angle In(E, N):

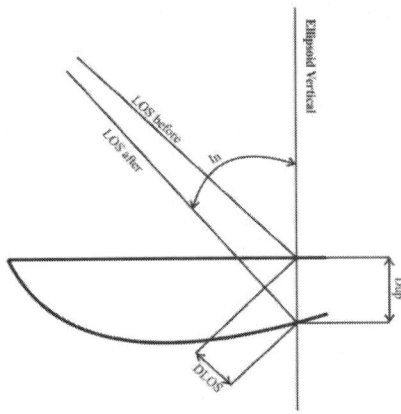

Figure 6. Relative geometry of the true vertical deformation and the deformation provided by InSAR.

$$\Phi_{\text{LOS}}(E, N) = \Phi_{\text{dU}}(E, N) \cdot \cos(\text{In}(E, N))$$

(2)

In order to determine the differential vertical displacements from LOS projection displacements, the value of the incidence angle for each target pixel was required. The incidence angle computation procedure was based on satellite trajectory data and the position of the target. Initially, for every target pixel (i, j) the zero - Doppler position of the space born SAR sensor had to be computed. This was achieved through signal processing applied on the "master" (or "reference") image. Third degree polynomials were fitted with Least Squares to the known satellite position vectors $\mathbf{r}(t)$ derived by ERS 1/2 operational orbits provided in the header file of every SAR image. These expressions simply provide the satellite position vectors in the orbit's terrestrial geocentric reference frame as a function of time. Three polynomials were derived, one for every coordinate X, Y and Z. Exactly the same procedure was applied for the satellite velocity vector $\dot{\mathbf{r}}(t)$ and three additional equations were also obtained. Therefore, for every single target (i, j) the following procedure was followed:

1. The map projection coordinates of the target were converted to geocentric Cartesian coordinates in the geodetic terrestrial reference frame in which the satellite orbits were provided (in this particular case from HGRS 87 map coordinates to ITRF 96 geocentric Cartesian coordinates).

2. The mean Doppler frequency shift was computed by the CNES DIAPASON software and was assumed to be the same for every single pixel target. The Doppler frequency shift $f(i,j)$ was expressed as a function of the satellite position, the satellite velocity vectors and the target position $\mathbf{r}(i, j)$, by the following equation (λ denotes the SAR sensor wavelength):

$$f(i,j) = \frac{2(r(i,j) - r(t_i)) \cdot \dot{r}(i,j)}{\lambda |r(i,j) - r(t_i)|}$$

(3)

3. A total of seven equations were accumulated, and an equal number of unknowns was introduced, three for the satellite position vector, three for the satellite velocity vector and one for the time t_i. Hence, a non linear seven-equation system was created for the estimation of the seven unknowns. The system was linearised with Taylor series expansion and solved iteratively.

4. Knowing the satellite and target position vectors, the unitary LOS vector could be calculated simply from the following vector equation:

$$\mathbf{LOS}\,(i,j) = \frac{r(i,j) - r(t_i)}{|r(i,j) - r(t_i)|}$$

(4)

5. The target position ellipsoidal coordinates $\varphi_{i,j}$, $\lambda_{i,j}$ were then calculated on the same geodetic terrestrial frame, which was used to express the orbits and the target coordinates in the previous step.

6. Knowing the target's latitude and longitude $\varphi_{i,j}$, $\lambda_{i,j}$ the LOS vector components were transformed to the local geodetic reference system (delta north - DN, delta east - DE, delta up -DU) by means of a rotation matrix:

$$\begin{bmatrix} DX(t) \\ DY(t) \\ DZ(t) \end{bmatrix} R\,(\varphi_{i,j}, \lambda_{i,j}) = \begin{bmatrix} DN(t) \\ DE(t) \\ DU(t) \end{bmatrix}$$

(5)

7. The third component of the **LOS** vector as expressed in the local geodetic reference system is actually the direction cosine for the "up" axis of the system, and consequently the cosine of the incidence angle In. Thus the incidence angle can be derived as:

In=arctan (DU)

(6)

Stacking

In the framework of this study and due to the fact that reliable verification data were available through the leveling survey, it was possible to evaluate the advantage in using a mean stacked interferogram instead of using only one, that is the "highest-quality" (most coherent) interferogram. For this purpose the sixteen "tilted" and "shifted" unwrapped interferograms were stacked to derive a mean temporal deformation field. This technique produced an image S(i, j) defined as: S(i, j) = mean(I_1(i, j), I_2 (i, j),... I_n(i, j)), where n represents the number of the available interferograms and I_m(i, j) the unwrapped interferometric phase of the mth interferogram at pixel location (i,j). Consequently the produced interferogram depicting the mean deformation field, was released from high and intermediate frequencies [21], which corresponded to non-earthquake related interferometric disturbances (Figure 7).

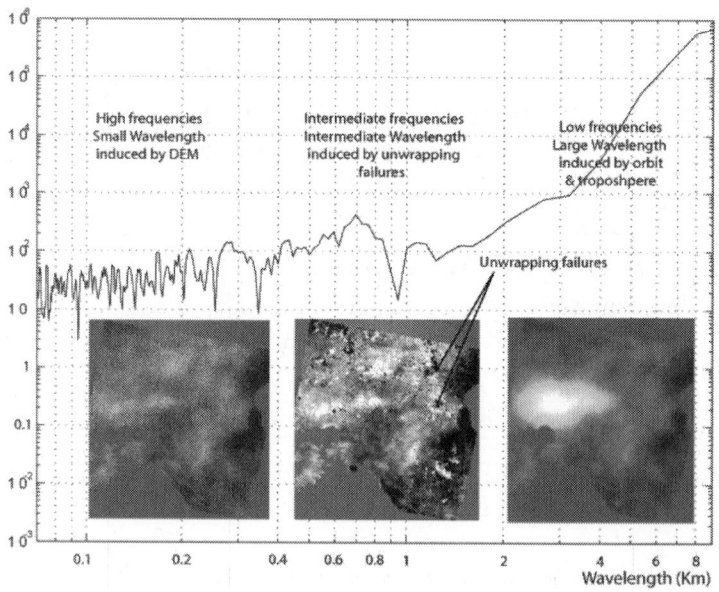

Figure 7. Spectral density of the stacked interferogram. Low frequencies prevail.

It should be mentioned that at this stage alternative stacking methods were implemented as well. They comprised of the formation of A) a weighted mean stacked interferogram, using as weights the pixel coherence values of each contributing interferogram, B) a maximum coherence stacked product, on which each phase pixel value stems from the interferogram with the highest corresponding coherence pixel value and C) a windowed maximum coherence stacked product; here each phase pixel value stems from the interferogram with the highest mean coherence value, calculated inside a 3 by 3 pixels window, centered on the pixel of interest. As is shown in section 4, the above methods returned very similar results compared to the mean stacked approach.

Geodetic Reference System Conversion

As mentioned the unwrapped interferometric calculations were referring to a UTM map projection on the ED 50 Greek Datum. In contrast the coordinates of the height references were expressed in the HGRS 87 reference system, using the Transverse Mercator map projection on the GRS 80 ellipsoid. To overcome this incompatibility the initial interferograms were converted to HGRS 87 projection system as follows:

1. The ED 50 UTM map coordinates (Eastings and Northings - E, N) were converted to ED 50 ellipsoidal coordinates (latitude and longitude -φ, λ), assigning to each pixel the corresponding orthometric height (H) derived from the input DEM.
2. The orthometric heights were converted to geometric ones (h), by implementing a constant additive geoid undulation value (N) for the entire area of interest, since the geoid in this area is relatively "flat" exhibiting a very low gradient. This value was obtained by the Ohio State University OSU 91 Geoid Model, and was recomputed for ED 50.
3. The ED 50 ellipsoidal coordinates were converted to ED 50 Cartesian coordinates (X, Y, Z).
4. Subsequently, the ED 50 geocentric Cartesian coordinates were converted to HGRS 87 geocentric ones, assuming only a parallel shift between the two systems. The latter assumption was expected to successfully provide the conversion due to the small size of the area of interest.
5. Then, the HGRS 87 geocentric Cartesian coordinates were translated to HGRS 87 ellipsoidal (φ, λ) coordinates.

6. Ultimately, the HGRS 87 ellipsoidal (φ, λ) coordinates were converted to HGRS 87 Transverse Mercator projection coordinates (E, N).

Differential Vertical Displacement Modeling

Thorough examination of the unwrapped (stacked and/or "highest-quality") interferograms, exhibited the presence of "local" phase anomalies in certain areas extending from one to several pixels. The phase values in these pixels deviated from the prevailing values in the surrounding region. These anomalies were survived the filtering procedure described in section 3.1. It is beyond the scope of this paper to explore the origin of such phase "residuals", but it could be assumed that they stemmed from local temporal decorrelation. It was also observed that the areas affected by these anomalies, presented significantly low coherence values and therefore they should be excluded.

Because the tectonic deformations observed were characterized by low phase gradient and spatial continuity, it was decided to proceed with a phase smoothing operation, by fitting (with the Weighted Least Squares method) a 3D-mathematical surface to the unwrapped interferometric phases Φ_{dU}(E, N). After a series of adjustments, a successful fit according to the chi-square (χ^2) test was achieved, using the value of 6 mm as a-priori standard deviation for the observations. By the application of the error propagation law (given the estimated model parameters and their a-posteriori standard deviation values), it was concluded that the 3D-mathematical surface would provide the vertical deformation estimate for each target pixel (E, N), with an estimated a-priori deviation not higher than 0.2 mm. In order to ensure that the mathematical model represents the best fit to the displacement pattern observed, the most general form of mth degree surface was tested:

$$\Phi_{dU}\,(E, N) = -\left(\begin{array}{l} a_0 + a_{E_1} E^1 + a_{E_2} E^2 + \ldots a_{E_m} E^m + \\ a_{N_1} N^1 + a_{N_2} N^2 + \ldots a_{N_m} N^m + a_{EN} E^1 N^1 + \\ a_{E_1 N_{(m-1)}} E^1 N^{(m-1)} + a_{E_2 N_{(m-2)}} E^2 N^{(m-2)} + \ldots a_{E_{(m-1)} N_1} E^{(m-1)} N^1 \end{array} \right)$$

$$(7)$$

After several runs, it was determined that a polynomial surface with degree higher than third would be redundant, as it was not offering any further improvement in terms of a-posteriori variance and measurement residuals. All higher degree coefficients were close to zero. The produced surface is presented in Figure 8. A Gaussian 3D surface was also tested; however this model was far less successful, mainly due to the absence of axial symmetry of the deformation pattern.

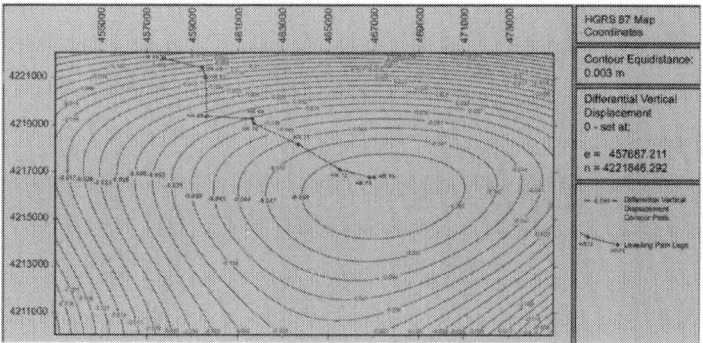

Figure 8. Differential vertical displacement model using a third degree mathematical surface.

RESULTS

Based on the 3D surface model produced, it became possible to extract a profile section of the InSAR vertical differential displacements along the leveling traverse. For this an origin had to be defined, and this was decided to be the height reference HR 65. Consequently, its displacement was set to zero. All other vertical displacements were provided in relevance to HR65. Profile data for InSAR and leveling data are presented in Figure 9.

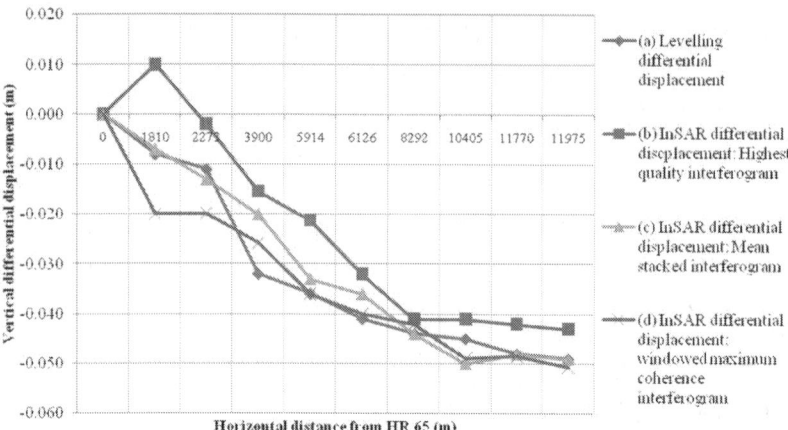

Figure 9. Differential vertical deformation profiles derived by the, (a) conventional terrestrial surveying, (b) single "highest quality" interferogram, (c) mean stacked interferogram, (d) windowed maximum coherence interferogram. HR65 indicates the starting point of leveling.

Examining the profiles illustrated in Figure 9, it can be concluded that no major differences occur between the differential vertical displacements as obtained by InSAR and leveling. There appears to be an agreement between the two profiles with respect to the gradient of the vertical displacement. Also there is no evidence of any systematic deviation between them. Moreover the profile corresponding to the mean stacked interferogram shows a better agreement with the leveling data. The vertical displacement differences between the leveling data and the interferometric data using the "highest quality" interferogram range from 3 mm up to 1.8 cm. The average difference value between the two data sets is 9.5 mm and the standard deviation equals 5.5 mm. On the contrary, when the mean stacked interferogram is compared with the leveling data, the above discrepancies are reduced by a factor of six. Indeed, the average difference between the two data sets is reduced down to 1.5 mm, whereas the standard deviation is of the order of 4.8 mm.

Table 1 outlines the average vertical displacement difference between the leveling data and the interferometric data for the various interferometric approaches used. The study of the table shows that the mean stacked product is preferred against the other interferograms, as it fits precisely the leveling data. Also its estimation entails less computational complexity. It should be mentioned though, that there are no major differences between the various stacking methods. However, significant improvement was achieved when moving from the single most coherent interferogram to any of the stacked products.

Table 1. Average and standard deviation of the vertical displacement differences between the leveling data and the InSAR methods.

	Highest quality interferogram	Mean stacked interferogram	Weighted mean stacked interferogram	Maximum coherence stacked interferogram	Windowed maximum coherence stacked interferogram
Average difference (m)	−0,0096	−0,0016	−0,0030	0,0047	0,0020
Standard deviation (m)	0,0056	0,0048	0,0055	0,0150	0,0056

CONCLUSIONS – DISCUSSION

This research focused on rendering compatible and comparable the InSAR derived displacements, related to the September 7, 1999 Athens earthquake, with leveling survey data. Towards this goal a processing chain was implemented, encompassing an algorithm for orbit and atmospheric disturbances removal, and a methodology for transforming the LOS deformation vector to the true vertical deformation vector. The proposed method used a mean stacked interferogram to get a more consistent representation of the displacement pattern. Finally, an agreement between the deformation values originating from InSAR data with the ones derived from leveling survey data was demonstrated. Only minor discrepancies were identified between the two.

These small differences may be attributed to several types of error sources, such as 1) SAR sensor noise, radiometric instabilities and system aging, 2) surface subsidence model deviations, 3) remaining orbital phase "ramps", 4) remaining tropospheric artifacts, 5) unwrapping errors, 6) temporal decorrelation effects, and 7) DEM errors. The possibility for a-seismic deformations in the period 1984-1998 could be also considered as a possible contributor to the relative subsidence profile differences. However, this a-seismic tectonic deformation, if it exists, remains unaccounted for, due to the absence of InSAR calculations in that period. The above-mentioned factors, may contribute to the observed total error of the derived relative subsidence values. However, the combined influence of the first three factors is considered to be essentially ignorable, taking into account the orders of magnitude of the resulting relative vertical displacement differences. Moreover, unwrapping errors computed by rewrapping the unwrapped interferogram, were assessed to be to an acceptable level in the area of interest. Therefore factors 6 and 7 namely temporal decorrelation and DEM errors, seem to be the most crucial parameters resulting in InSAR subsidence profile deviations. Temporal decorrelation could not be computed but must be considered as the major contributor to the spatially uncorrelated component of the residuals arising from the Least Squares approximation of the polynomial surface.

However as shown in Figure 9, the influence of all disturbing factors described previously, was effectively reduced by using a mean stacked and noise-free interferogram. Moreover the suggested tilting and shifting procedure, introduced in [21], for removing orbital and tropospheric fringes has performed effectively. Hence, the earthquake induced subsidence pattern seemed to be successfully represented by the proposed model.

As far as the terrestrial surveying derived relative subsidence profiles are concerned, the estimation accuracy was much simpler and more explicit. The leveling data accuracy was estimated to lie in the range from 4 mm to 8 mm, in relative heights between successive height benchmarks. With the above estimations it becomes clear that the deviation of the two relative subsidence profiles (cases (a) and (c) inFigure 9), fall entirely within the confidence interval defined for the leveling data. It can be also concluded that the simple polynomial surface modeling of the subsidence field, may be regarded as an effective method to overcome the remaining temporal decorrelation effects and other sources of noise, by exploiting the extremely high degrees of freedom associated with the Least Squares approximation of mathematical models. Finally, a case specific conclusion of geophysical interest can be drawn for the study area. This refers to the fact that no detectable significant vertical displacements have occurred during the period 1984-1998, for which InSAR interferometric measurements were not available.

We are grateful to the European Space Agency (ESA) for providing us the ERS1/2 data, in the framework of the ESA-GREECE AO project 1489OD/11-2003/72.

REFERENCES

1. Kontoes, C.; Elias, P.; Sykioti, O.; Briole, P.; Remy, D.; Sachpazi, M.; Veis, G.; Kotsis, I. Displacement Field Mapping and Fault Modeling of the Mw = 5.9, September 7, 1999 Athens Earthquake based on ERS-2 Satellite RADAR Interferometry. *Geophysical Research Letters* **2000**,*27*, 3989–3992.

2. Roumelioti, Z; Dreger, D.; Kiratzi, A.; Theodoulidis, N. Slip distribution of the 7 September 1999 Athens earthquake inferred from an empirical Green's function study. *Bulletin of the Seismological Society of America* **2003**, *93*, 775–782.

3. Baumont, D.; Courboulex, F.; Scotti, O.; Melis, N.S.; Stavrakakis, G. Slip distribution of the M-w 5.9, 1999 Athens earthquake inverted from regional seismological data. *Geophysical Research Leters* **2002**, *29*, 1720.

4. Papadimitriou, P.; Voulgaris, N.; Kassaras, I.; Kaviris, G.; Delibasis, N.; Makropoulos, K. The M_w=6.0, 7 September 1999 Athens earthquake. *Natural Hazards* **2002**, *27*, 15–33.

5. Sargeant, S.L.; Burton, P.W.; Douglas, A.; Evans, J.R. The source mechanism of the Athens earthquake, September 7, 19999, estimated from P seismograms recorded at long range. *Natural Hazards* **2002**, *27*, 33–45.

6. Pavlides, S.B.; Papadopoulos, G.; Ganas, A. The fault that caused the Athens September 1999 Ms=5.9 earthquake: Field observations. *Natural Hazards* **2002**, *27*, 61–84.

7. Goldsworthy, M.; Jackson, J.; Hains, J. The continuity of the active fault systems in Greece.*Geophysical Journal International* **2002**, *148*, 596–618.

8. Bouckovalas, G.D.; Kouretzis, G.P. Stiff soil amplification effects in the 7 September 1999 Athens (Greece) earthquake. *Soil Dynamics and Earthquake Engineering* **2001**, *21*, 671–687.

9. Eftaxias, K.; Kapiris, P.; Polygiannakis, J.; Borgis, N.; Kopanas, J.; Antonopoulos, G.; Peratzakis, A.; Hadjicontis, V. Signature of pending earthquake from electromagnetic anomalies. *Geophysical Research Letters* **2001**, *28*, 3321–3324.

10. Odijk, D.; Kenselaar, F.; Hanssen, R. Integration of Leveling and InSAR Data For Land Subsidence Monitoring. Proc. 11th International FIG Symposium on Deformation Measurements 2003, Commission 6. Santorini Island, Greece; p. 8.

11. Zhou, Y.; Stein, A.; Molenaar, M. Integrating Interferometric SAR data with Leveling Measurements of Land Subsidence Using Geostatistics. *International Journal of Remote Sensing* **2003**, *24*, 3547–3564.

12. Ge, L.; Chang, H.C.; Janssen, V.; Rizos, C. The Integration of GPS, radar Intrerferometry and GIS for Ground Deformation Monitoring. Proc. Int. Symposium on GPS/GNSS 2003, Tokyo, Japan; pp. 465–472.

13. Liu, G.; Chen, Y.; Ding, X.; Li, Z.; Li, Z.W. Monitoring Ground Settlement in Hong Kong with Satellite SAR Interferometry. *Proc. FIG XXII International Congress Washington 5* **2002**, *JS17*, 12.

14. Massonet, D.; Feigl, K. Radar interferometry and its application to changes in the Earth's surface. *Reviews of Geophysics* **1998**, *36*, 441–500.

15. Balodimos, D. The development of a special trigonometric leveling technique. *Tecnika chronic* **1979**, *3*. (in Greek).

16. Deltsidis, P.; Saridakis, M. Displacement determination with terrestrial and satellite methods. Diploma Thesis, Dionysos Satellite Observatory, National Technical University of Athens, (in Greek). 2001.

17. Prit, M.D. Comparison of path-following and least-squares phase unwrapping algorithms. *IGARSS* **1997**, *2*, 872–875.

18. Zebker, H.A.; Lu, Y. Phase unwrapping algorithms for radar interferometry: Residue-cut, least squares, and synthesis algorithms. *Journal of the Optical Society of America* **1998**, *15*, 586–598.

19. Pritt, M.D. Congruence in least-squares phase unwrapping. *IGARSS* **1997**, *2*, 875–877.

20. Ghiglia, D.C.; Romero, L.A. Minimum LP-norm two-dimensional phase unwrapping. *Journal of the Optical Society of America* **1996**, *13*, 1999–2013.

21. Elias, P.; Kontoes, C.; Sykioti, O.; Avallone, A.; Briole, P.; Paradissis, D. A method for minimizing low frequency and unwrapping artefacts from interferometric calculations. *International Journal of Remote Sensing* **2006**, *27*, 3079–3086.

CITATION

Ioannis Kotsis, Charalabos Kontoes, Dimitrios Paradissis, Spyros Karamitsos, Panagiotis Elias and Ioannis Papoutsis, A Methodology to Validate the InSAR Derived Displacement Field of the September 7th, 1999 Athens Earthquake Using Terrestrial Surveying. Improvement of the Assessed Deformation Field by Interferometric Stacking, doi:10.3390/s8074119

CHAPTER 8

Landslide Investigation with Remote Sensing and Sensor Network: From Susceptibility Mapping and Scaled-down Simulation towards *in situ* Sensor Network Design

Gang Qiao [1], Ping Lu [1],* Marco Scaioni [1], Shuying Xu [1], Xiaohua Tong [1], Tiantian Feng [1], Hangbin Wu [1], Wen Chen [1], Yixiang Tian [1], Weian Wang [1] and Rongxing Li [1,2]

[1]Center for Spatial Information Science and Sustainable Development Applications, College of Surveying and Geo-Informatics, Tongji University, Shanghai 200092, China;
[2]Mapping and GIS Lab, The Ohio State University, Columbus, OH 43210, USA

ABSTRACT

This paper presents an integrated approach to landslide research based on remote sensing and sensor networks. This approach is composed of three important parts: (i) landslide susceptibility mapping using remote-sensing techniques for susceptible determination of landslide spots; (ii) scaled-down landslide simulation experiments for validation of sensor network for landslide monitoring, and (iii) *in situ* sensor network deployment for intensified landslide monitoring. The study site is the Taziping landslide located in Hongkou Town (Sichuan, China). The landslide features generated by landslides triggered by the 2008 Wenchuan Earthquake were first extracted by means of object-oriented methods from the remote-sensing images before and after the landslides events.

On the basis of correlations derived between spatial distribution of landslides and control factors, the landslide susceptibility mapping was carried out using the Artificial Neural Network (ANN) technique. Then the Taziping landslide, located in the above mentioned study area, was taken as an example to design and implement a scaled-down landslide simulation platform in Tongji University (Shanghai, China). The landslide monitoring sensors were carefully investigated and deployed for rainfall induced landslide simulation experiments. Finally, outcomes from the simulation experiments were adopted and employed to design the future *in situ* sensor network in Taziping landslide site where the sensor deployment is being implemented.

INTRODUCTION

Landslides are major geo-hazards heavily impacting many regions of the world in terms of human lives and economic losses [1]. The large magnitude of natural forces that are involved in landslides makes actions of mitigation or prevention unfeasible, with exceptions for small occurrences or under particular conditions. According to [2], on the basis of methods employed landslide research can be classified into theoretical, analytical and numerical studies along with laboratory experiments, field investigations, monitoring and inventory mapping, as well as GIS/Remote Sensing application.

Landslide assessment, susceptibility mapping, monitoring (*in situ* and remotely) and early warning systems all play an important role in landslide investigations thereby being directly related to disaster reduction and hazard mitigation. Remote sensing, especially from high-resolution satellite imagery is gaining importance in landslide investigation due to its wide coverage [3]. With its increasing spatial and temporal resolution, remote sensing has been widely adopted in landslide mapping for rapid response and recovery after hazard occurrence by government agencies as well as research community [4]. When data are available, the comparison of images collected before and after the event gives relevant support to landslide recognition [5,6]. Apart from its use in emergency response, when incorporated with other vector data the landslide inventory data from landslide mapping results and historical records can be employed as a database for susceptibility mapping using GIS and statistical approaches [7–9].

Landslide susceptibility mapping is an attempt to derive spatial variation of area-based slope failure probability or instability at a regional scale. This is based on a number of factors categorized into (a) preparatory factors such as lithology and geomorphology; and (b) triggering factors such as seismicity, rainfall, land cover, and anthropogenic causes. This technique is a fundamental tool for planners, administrators, and emergency managers, who are responsible for identifying hazard zones [10,11].

As another important component in landslide research, landslide monitoring has usually been regarded as a point-based and site-specific slope stability analysis [12]. In a broad sense, the term "monitoring" refers to the observations over time of any change in the area of interest caused by the landslide process. This includes either qualitative properties (vegetation cover, urbanization, land use, surface processes, and the like) or geometric aspects (deformation monitoring) [13]. Monitoring is, on one hand, a fundamental tool for understanding and modeling all geo-processes related to landslide dynamics. On the other hand, the integration of models and observations can lead to a greater opportunity for predicting partial or full slope failures. Monitoring also plays a major role in early-warning systems, where it should provide timely and reliable information to evaluate the present risk. The closer is the relationship between monitoring and modeling, the smaller is the uncertainty of forecasted events. Indeed, both tasks can mutually benefit through the assimilation of observation data and empirical or physical models for the description of landslide behavior [14,15]. On one hand, dense and accurate observations can improve models. On the other, better models will help tailor the monitoring system to become more efficient in understanding geo-processes. This can be achieved, for example, by deployment of new sensors or by tuning the data acquisition rate according to the observed dynamics. In the latest generation of landslide monitoring systems, this capability to change the sensor parameters can be accomplished almost in an automatic way ("smart networks"). This, however, is a perspective of future development in this domain.

Landslides are complex geo-hazards requiring multiple observations to be investigated. For this reason, a large number of sensors could be installed as necessitated by the specific properties of each site. The term "sensor" is intended as in [16]. The logical connection between sensors is usually referred to as a "network", within which sensors are the "nodes". These nodes are implemented in a physical way using wired or wireless communication infrastructures [17] and different data transmission protocols [18]. In general, a Sensor Network (hereafter "SN") can comprehend the different categories of sensors that are in use for gathering information about the underground layers, about the topographic of the slope surface, and about meteorological conditions. Usually these data inputs need to be updated at high-to-medium level of frequency (*i.e.*, from minutes to within a few days). On the other hand, some low-frequency data can be useful for the completion of the observation budget. Major categories of sensors include [19]:

- Geotechnical sensors to gather observations from surface and sub-surface;
- Geophysical sensors focusing on the underground layers;
- Surveying techniques (including the Global Navigation Satellite System, or GNSS—see [20]) which provide point-wise deformation measurements;
- Remote-sensing techniques (photogrammetry, laser scanning, and Interferometric Synthetic Aperture Radar (InSAR)—see [21–24]) that can offer "area-based" observations from different platforms (spaceborne, airborne and ground-based); and
- Environmental sensors, which can cover local (e.g., whether stations), regional (e.g., meteorological radars), and wide-areas (*i.e.*, meteorological satellites).

Monitoring should be complemented by up-to-date information on triggering factors, such as earthquakes [25] or anthropogenic activities [13,26] other than rainfall. The definition of a general methodology [27] for designing of a SN for monitoring landslides is a challenging issue that depends strictly on the geological processes involved [28], on the extent of the area under investigation, on the specific features of each site, and on the available economic resources. A detailed study of the local geological properties and the inventory of past landslides in the same region, as well as the acquisition of available information on past slope failures would significantly improve the model. In addition, information concerning the surrounding region (human settlements, hydraulic networks, communication infrastructures) and important targets (such as river barrages, energy power-plants and the like) is fundamental for designing risk scenarios outlining the serious impacts that could be caused directly by the landslide or by other domino effect such as the Na-Tech disasters [29,30].

BACKGROUND OF THE RESEARCH

This research proposes a methodology for designing a SN, and its application to a landslide located in Sichuan province in southwestern China. Two important earthquakes ("Wenchuan" earthquake on 12 May 2008 [25] and "Ya'an" earthquake on 20 April 2013) have occurred in this area in the last five years, affecting different mountain locations. Both events were characterized by the triggering of several slope instability factors that, in the case of the 2008 earthquake, also coincided with intense rainfall resulting in floods and debris flows. In addition to the landslide runouts that occurred immediately after these earthquakes, many other slopes became unstable and represented a significant risk to man and property [31]. Considering the large area involved and the similarity between numerous unstable slopes (geometric properties, lithology, slope, soil, and ground coverage), the development of a unified, sustainable approach to landslide susceptibility mapping is quite important.

The key points of emphasis are: (i) the use of remote-sensing data together with some geo-referenced information from archives to determine the susceptible areas (see Section 5); and (ii) the construction of a scaled-down model of the slope to artificially reproduce the landslide process using a rainfall trigger (see Section 6). It is considered that future rainfall might be the highest probable triggering factor of new slope failures in the area [32]. The scaled-down model has two purposes. For one, it can be used to test the SN before it is deployed on the real slope [33]. In addition, it can provide observation results to be used in the process of numerical modeling of landslides.

RESEARCH AREA AND DATASET

The research area is located in Hongkou Town (Sichuan, China) within the transition zone of the Chengdu Plain and the Western Sichuan Plateau. The approximate extent of the area is 30 km² and is characterized by relatively high and rugged mountains with elevations ranging from 700 to 1,700 m above sea level as well as deeply incised valleys [34] (Figure 1). This area is situated in the Yingxiu-Beichuan fault zone, where earthquakes and heavy precipitation lead to geological hazards such as landslides and debris flows, for example after shock events of 2008 Wenchuan Earthquake [31]. The major mountain range in this area is the Longmenshan Mountains, mainly composed of conglomerate, carbonat, tuffaceous, arkose, limestone, and granite rock (*cf.*, National Geological Archives of China, NGAC). The 2008 Wenchuan Earthquake triggered a large number of landslides, rock avalanches, and debris flows. Some of the landslides formed natural dams in the rivers, causing the potential secondary hazard of subsequent flooding. One third of the estimated 88,000 casualties of the earthquake were thought to be caused by landslides alone [35].

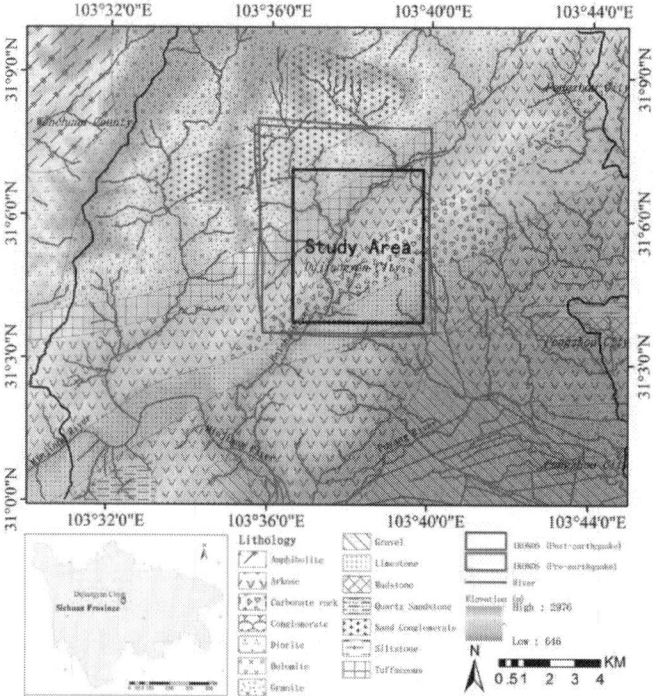

Figure 1. Study area and profile of the data coverage.

To map the landslides induced by the earthquake, two scenes of IKONOS imagery with almost the same coverage were used (Figure 1).The pre-event image was acquired on 14 September 2007, and the post-event one was taken one and a half months after the event on 28 June 2008.

In addition to this imagery, a DEM covering the research area having a grid resolution of 25 m × 25 m was provided by the National Geomatics Center of China (NGCC). Updated in 2011, this DEM is suitable to generate basic topographic information for landslide control factor analysis after the 2008 Wenchuan Earthquake. Updates of the road network and river data were obtained in 2011 from the Digital Line Graph (DLG) data offered by the NGCC. The lithology data and 2008-Wenchuan-Earthquake fault information were generated by the NGAC and the China Earthquake Networks Center (CENC), respectively.

METHODOLOGY

As shown in the flowchart (Figure 2), this study is composed of three parts: landslide susceptibility mapping for susceptible spot determination, a scaled-down landslide simulation experiment for SN test, and the *in situ* SN deployment of sensors for intensified observation of landslide activity. In the first part, landslides were extracted by means of an object-oriented method using the pre- and post-event IKONOS images. Then landslide control factors (including aspect, elevation, slope, lithology, distance to nearest river, distance to nearest road, and seismic intensity) were derived from the DEM and other vector data as listed in Table 1. The relationships between the distribution of the extracted landslides and the control factors were analyzed, and a landslide susceptibility map was generated based on Artificial Neural Network (ANN) to determine any susceptible spots (see Section 5 for details). In the second part, the Taziping landslide was selected as an example to design and implement a scaled-down landslide simulation platform at Tongji University (Shanghai, China). The landslide monitoring sensors were carefully investigated and deployed for this landslide simulation experiment triggered by rainfall, and conclusions concerning SN deployment were drawn. In the final part, current landslide prevention infrastructure and monitoring sensors were first introduced and discussed, and finally with the considerations from the simulation experiment, suggestions for future *in situ* SN deployment in Taziping were given.

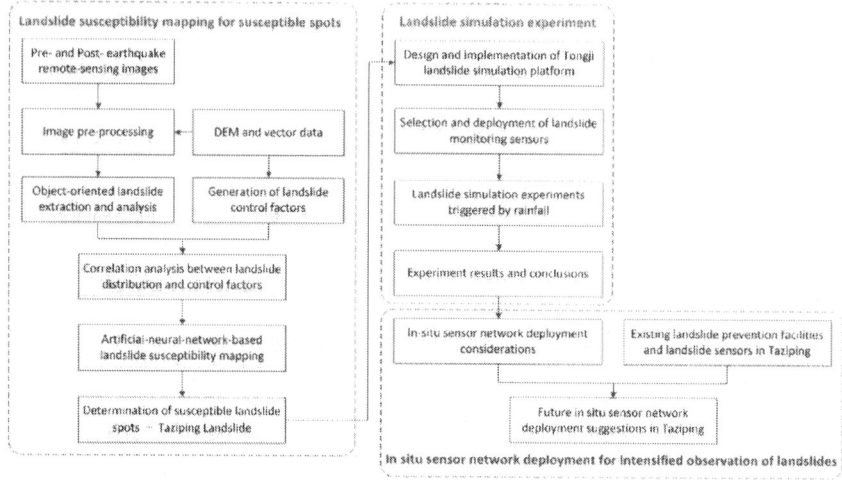

Figure 2. Flowchart of the research approach.

Table 1. Selected information on the IKONOS images and vector data used in the study.

Item	IKONOS Images		Vector Data				
	Pre-earthquake	Post-earthquake	DEM	Road	River	Lithology	Fault
Date	2007.9.14 12:12 pm	2008.6.28 12:02 pm	2011	2011	2011	2008	2008
Scale	NA	NA	1:50,000	1:50,000	1:50,000	1:500,000	NA
Resolution	1.0 m	1.0 m	25.0 m	NA	NA	NA	NA
Source	GeoEye Company (now DigitalGlobe)		NGCC			NGAC	CENC

LANDSLIDE SUSCEPTIBILITY MAPPING

Landslide Extraction Based on Pre- and Post- Earthquake IKONOS Images

The IKONOS images were first enhanced, rectified and geo-referenced [36] based on the corresponding points obtained from vector data such as road networks, rivers, and land-use maps. Then an object-oriented approach [37,38] was employed to extract the earthquake-induced landslides. Since such geo-hazards usually occur on higher slopes and thus destroy the vegetation-cover, for most of the rugged mountains in this area any changes in vegetation-cover between the pre and post-event images could be viewed as landslide indicators [39,40]. A Vegetation Damage Index (VDI) is hence defined here and used for landslide recognition:

$$VDI=NDVI_{Pre}-NDVI_{Post} \qquad (1)$$

where $NDVI_{Pre}$ is the Normalized Difference Vegetation Index (NDVI) [41] in pre-event image, and $NDVI_{Post}$ is the corresponding parameter in the post-event image.

A multi-resolution segmentation algorithm [37,42,43] was applied to the post-event IKONOS image to identify the homogeneous regions (namely "image objects") and the $NDVI_{Post}$ value for each object was computed. Then the image objects were overlaid with the pre-event IKONOS image to obtain the $NDVI_{Pre}$ values. An empirical threshold for the VDI of 0.36 was obtained after several trials based on visual inspection of resulting VDI map compared with observed landslides in the post-event IKONOS image in the research area. This value was used for the extraction of landslides from both images. Results were visually checked, and any errors (e.g., features such as roads sheltered by trees in the pre-event image and revealed in the post-event image, and some of the landslides with smaller VDI due to lower contrast) were edited to ensure the extraction accuracy by minimizing the number and area of the declassified landslides.

Figure 3a shows the extracted landslides overlaid on the post-event image as well as the river layer. In general, the landslide area shows dark brown contrast where it is compared with the nearby surroundings in remotely sensed images. The path of debris movement could be discerned easily, as the landslide resulted in newly denuded vegetation in comparison with the pre-event image [44]. The landslides were mainly situated on the hanging wall of the precipitous mountain ridges and on into the steep valley sides of the Baisha River and its tributaries. Most of the landslides took place in the western part of the study area with some very large ones in the southwestern section while only a few occurred in the eastern part where the terrain relief is relatively flat (see also Figure 4c). Figure 3b illustrates a detailed landslide pattern by the side of a village area with a background of the pre- and post- event images to demonstrate the apparent changes caused by debris movement. Here it can easily be observed that some buildings were damaged, and a major trunk road was buried and blocked by a large landslide.

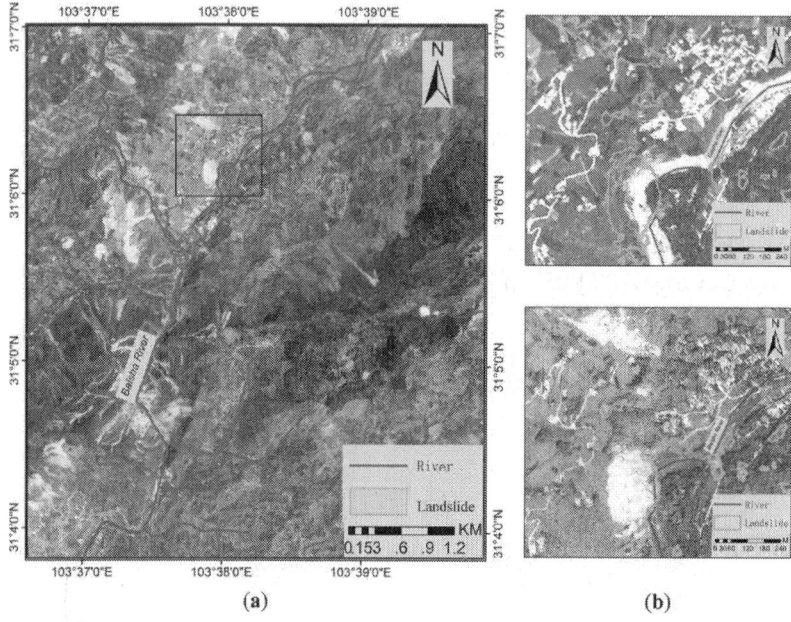

(a)   (b)

Figure 3. The landslides extracted from the two IKONOS images: **(a)** Landslide distribution overlaid on the post-event IKONOS and the river maps; and **(b)** A detailed landslide pattern (the outlined area in (a)) with a background of pre-event (upper) and post-event (lower) IKONOS images.

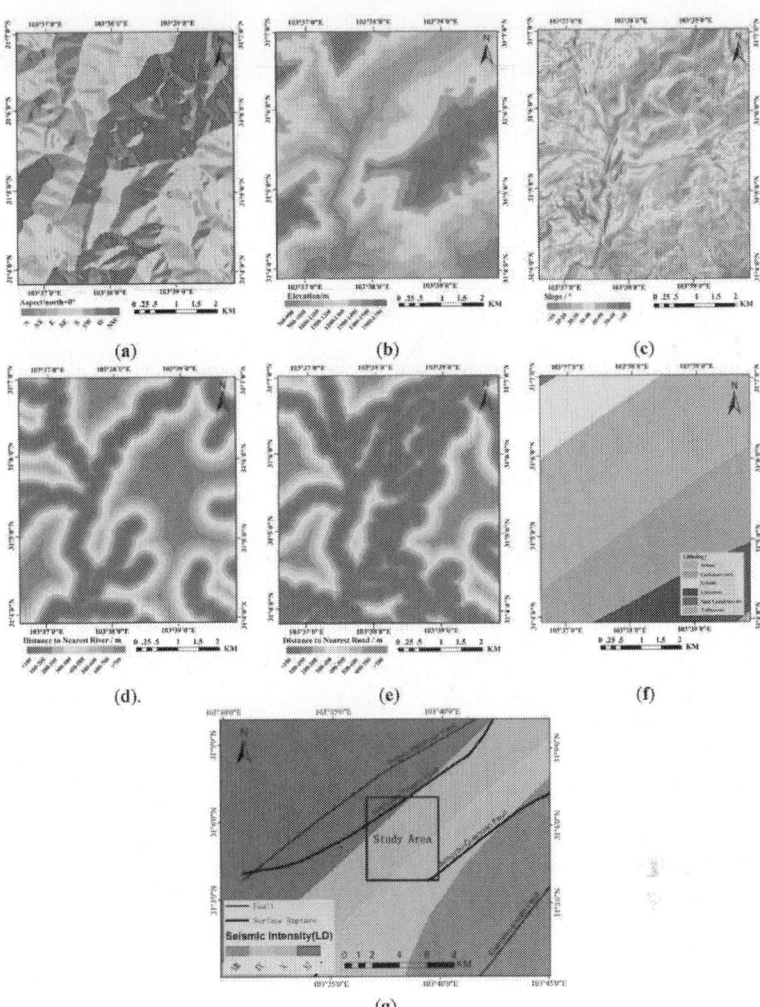

Figure 4. The topographical (**a–e**), geological (**f**), and seismic (**g**) factors for landslides in the study area. (a) Aspect; (b) Elevation; (c) Slope; (d) Distance to nearest river; (e) Distance to nearest road; (f) Lithology; (g) Seismic intensity.

Table 2 shows the landslide numbers and area extracted from the remote-sensing images. For the IKONOS images having a 1.0 m spatial resolution, landslides with an area of less than 10 m² could sometimes be identified, but errors could also be introduced in this case, so they were removed from the statistics. In total, the extracted landslides cover a surface area of about 3.25 km², more than 10% of the research area, indicating severe damage by the landslides. The dimensions of individual landslide ranged from about 10 m² to more than 10,000 m². Most of the selected 740 landslides are with an area of 100–500 m², making almost half of the total number (41.62%), while a relatively small number (40) of large landslides (more than 10,000 m² in area) make the highest percentage (82.05%).

Table 2. Extracted landslide statistics in the research area.

Landslide Dimension (m²)	Landslide Area		Landslide Number	
	Area (m²)	Percentage (%)	Number	Percentage (%)
10–50	1,718.76	0.05	58	7.84
50–100	6,659.45	0.20	88	11.89
100–500	78,931.20	2.43	308	41.62
500–1,000	62,232.23	1.91	90	12.16
1,000–5,000	271,108.12	8.34	133	17.97
5,000–10,000	163,055.36	5.01	23	3.11
>10,000	2,667,869.77	82.05	40	5.41
Total	3,251,574.89	100.00	740	100.00

Correlation between Control Factors and Landslide Distribution

Generation of Control Factors

The control factors mentioned above are generally divided into topographical and geological factors [45,46]. In addition, seismic factors also are taken into account for earthquake-triggered landslides [47,48]. In this study, all three types of factors were considered. Topographical factors included aspect, elevation, slope, distance to nearest road, and distance to nearest river. The geological factor adopted was lithology. The seismic factor was the intensity of the 2008 Wenchuan Earthquake. It should be noted that the distance from the epicenter (approx. 30 km) was not used for this analysis because of the relatively small extent (approx. 30 km²) of the study area. This parameter would give a similar contribution in the whole region.

Figure 4 shows the landslide control factors in the study area: topographical (Figure 4a–e), geological (Figure 4f, lithology), and seismic (Figure 4g, seismic intensity). For the topographical factors, the aspect, elevation, and slope were derived from the DEM, and both the distance to nearest river and distance to nearest road were derived from the DLG data provided by NGCC. The lithology was generated from the 1:500,000 geological maps provided by the NGAC, and the 2008 Wenchuan Earthquake seismic intensity map was provided by the CENC. 5.2.2. Correlation Analysis between the Control Factors and Landslide Spatial Distribution

In order to analyze the correlation between control factors and landslide spatial distribution, the former were first classified into different grades according to their specific characteristics and data ranges (seeTable 3). The aspect factor was divided based on different direction classes with an amplitude of 45°. The elevation was classified based on 100-m intervals within the height range in the study area, while for the slope factor, classes of 10° were selected. For the distance-to-nearest-river/road factors, a 100-m interval was applied within the 700 m distance to obtain the grades. The lithology was classified on the basis of main rocks in the study area. The IX, X and XI seismic intensities (Richter scale) of the 2008 Wenchuan Earthquake were within the research area for further analysis.

Table 3. Detailed grades of the control factors used for correlation analysis.

Grade	Aspect (°)	Elevation (m)	Slope (°)	Distance to Nearest River (m)	Distance to Nearest Road (m)	Lithology	Seismic Intensity
1	N	768–900	<10	<100	<100	Conglomerate	IX
2	NE	900–1,000	10–20	100–200	100–200	Carbonate Rock	X
3	E	1,000–1,100	20–30	200–300	200–300	Tuffaceous	XI
4	SE	1,100–1,200	30–40	300–400	300–400	Arkose	
5	S	1,200–1,300	40–50	400–500	400–500	Limestone	
6	SW	1,300–1,400	50–60	500–600	500–600	Granite	
7	W	1,400–1,500	60–68	600–700	600–700		
8	NW	1,500–1,706		>700	>700		

On the basis of this dataset, two parameters were determined for analysis of the correlation between landslides and control factors: (1) Landslide Occurrence Rate in All landslides (LORA) and (2) Landslide Occurrence Rate in a certain landslide Grade (LORG) [49]. The former can be defined as the ratio of the landslide area within a certain grade with respect to the total landslide area extracted from satellite images of the area. The latter is the ratio of the landslide area within a certain grade to the area of this grade in the research region. LORA represents the percentage of the landslide area in a certain grade, while LORG denotes the possibility of landslide occurrence for a certain grade within the study area.

Figure 5 shows the correlation between landslides and the control factors with the blue line representing the LORA and the red the LORG. For the aspect factor, the values of both LORA and LORG are similar in the research region, and the two indexes are higher in the NE, E, SE and SW than in other directions, revealing that there are more landslides occurring in these directions. For the elevation factor, the LORA value (landslide area) is larger in the elevation range of 900–1,400 m, but the LORG is relatively smooth for all the grades, indicating no obvious correlation in elevation factor. For the slope factor, the LORA first increases and then decreases with the summit at 30°–40°, indicating the larger landslide area within the middle of the slope angle range, while the LORG ascends along with the rising slope angle, showing clear dependency between the landslide possibility and the slope angle and that higher slope areas are more unstable for the occurrence of the earthquake-triggered landslides. For the distance-to-nearest-river/road factors, a similar trend could be viewed for the LORA lines, and both values decrease progressively as the distance increases, showing that more landslides occurred near the rivers or roads. For the LORG lines, the distance-to-nearest-river [50] value has a similar degression tendency, indicating that the landslides occurred easily in valleys with flowing rivers. The distance-to-nearest-road value shows an increasing and then decreasing curve with the maximum in a range of 200-to-500 m, which is most probably due to the fact that roads are built mainly within this distance to mountains having a slope of 30°–40°. For the lithology factor, the landslides occurred mostly in regions of arkose and carbonate rock, while granite and tuffaceous are rocks most vulnerable to earthquake-triggered landslides. As to the seismic intensity factor, the X degree takes most of the landslide area, while possibility of landslides increases with the seismic intensity from IX, X to XI. This is consistent with common knowledge.

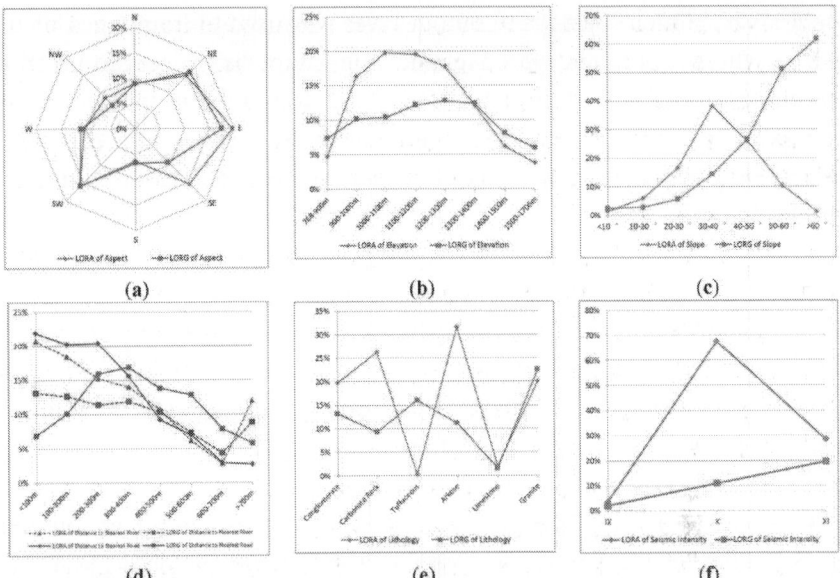

Figure 5. Relationships between landslides and the control factors: aspect, elevation, slope, distance to nearest river/road, lithology, and seismic intensity. (**a**) Aspect; (**b**) Elevation; (**c**) Slope; (**d**) Distance to nearest river/road; (**e**) Lithology; (**f**) Seismic intensity.

Landslide Susceptibility Mapping Based on an Artificial Neural Network

Landslide susceptibility mapping will provide an effective means for determining the susceptible regions for landslide monitoring. Landslide susceptibility mapping is commonly used in landslide research and prediction [51–53]. As a widely used non-linear method of prediction, the Artificial Neural Network (ANN), which is a generalization of mathematical models to simulate the human cognition and neural biology and thus to predict outputs from inputs [54], was employed in this research to derive the landslide susceptibility map. ANN has been recently used in landslide susceptibility mapping, and was proved to be an effective means for landslide prediction due to its capability to establish the complex and non-linear relationships between landslide distribution and the control factors [11].

In a typical landslide susceptibility mapping application, the ANN is usually composed of three steps: training, validation, and susceptibility mapping [55,56]. A back-propagation learning algorithm, consisting of input layer, hidden layers, and output layer was used to train the ANN in this research. The back-propagation algorithm is a procedure that minimizes the Sum of Squares for Error (SSE) in output layer by modifying the weights of ANN parameters to approximate the input sample and output sample. This multi-layer back-propagation algorithm was selected in this paper because it is one of the most commonly used ANN models that has the advantages of non-linear mapping, self-learning, and fault tolerance with respect to other models [57].

The landslides extracted from the IKONOS images were randomly divided into two groups, one used for training the ANN, and the other for accuracy assessment. The training sites were selected based on analysis of the topographical (aspect, elevation, slope, and distance to river/road), geology (lithology), and seismic (seismic intensity) landslide control factors. Based on the analysis described in the previous section, the LORG index of the control factors was used as the standard inputs for the network training due to its more reasonable evaluation compared with LORA, and the outputs of the training were the spatial distributed landslides extracted from satellite images. We tested different training methods, including the improved back-propagation algorithms and the numerical optimization based techniques, and found that the Levenberg-Marquardt approach featured the best performance in terms of Mean Square Error (MSE) [58,59], So it was selected in the back-propagation algorithm to calculate weights between the input layer and the hidden layer, as well as between the hidden layer and the output layer, by modifying the number of hidden layers and the learning rate [60] (see Table 4). The activation function in the hidden layer was set to tansig, and that in the output layer was set to purelin [58]. The input node (neuron number) was set to seven for the corresponding control factors and the output node (neuron number) was set to 1 for the susceptible value. The neuron number in the hidden layer was determined to be nine by evaluation of the training time, iterations and the resulting SSE (see Figure 6). 80% of the landslides were randomly selected as training data for the ANN, and the remainder was used as check points to validate the accuracy of the susceptibility mapping. Finally, with 58 iterations, an MSE (Mean Square Error) of 0.308 was achieved for the Levenberg-Marquardt method, and an accuracy of 95% could be reached for the assessment.

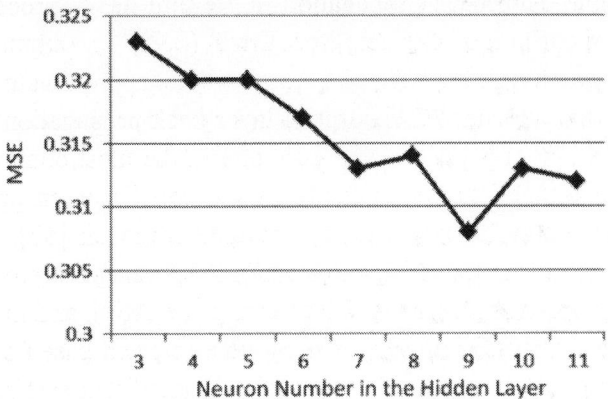

Figure 6. Mean Square Error (MSE) *vs.* neuron number in the hidden layer (Levenberg-Marquardt method).

Table 4. The Weights of the Artificial Neural Network (ANN) Parameters (Levenberg-Marquardt method).

Network Parameters	Weights
Learning Rate	0.05
Threshold of Residue	9.5238×10^{-4}, or SSE = 0.02
Error Increase Factor	1.04
Minimum Gradient	1×10^{-10}
Number of Maximum Training	2,000

The landslide susceptibility map derived from the ANN is shown in Figure 7, where the level of susceptibility was divided into five categories according to the output value r: very low ($r < 0$), low ($0 \leq r < 0.25$), moderate ($0.25 \leq r < 0.5$), high ($0.5 \leq r < 0.75$), and very high ($r \geq 0.75$). The overlay rate between the extracted landslides and the ones predicted by the susceptibility map is 89%, indicating the high reliability of the ANN approach.

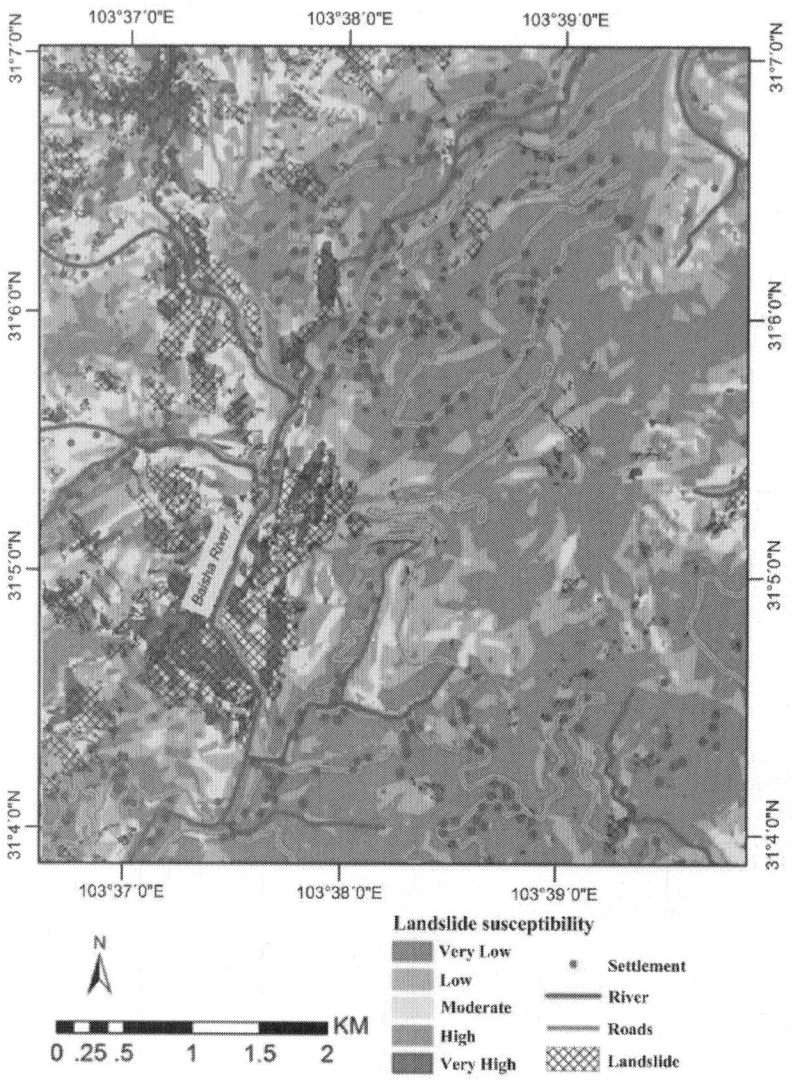

Figure 7. Landslide susceptibility map derived from Artificial Neural Network (ANN) (Levenberg-Marquardt method).

Figure 7 shows that the more landslide-prone regions are located mainly in the rugged mountainous areas having high slopes. The western and the northwestern areas are those most commonly susceptible to landslides, and there are relatively fewer settlements in these regions. One of the most susceptible spots, Taziping, was selected for landslide monitoring based on an *in situ* geo-sensor network because there is a river flowing nearby, and because there are relatively densely populated settlements in the area (Figure 7) which could be severely damaged if a landslide occurs. Before deploying sensors in the field of Sichuan, a scaled-down landslide simulation and monitoring system was constructed on the campus of Tongji University to test the usability and deployment strategy of these *in situ* sensors. The simulation experiment is detailed in the next section.

THE SCALED-DOWN LANDSLIDE SIMULATION

A simulation platform can be used as a first step in defining the set of *in situ* sensors to be adopted in the field. In addition, it can be used to understand the data interpretation and processing approach. For these purposes, a scaled-down landslide simulation platform (Figure 8) has been set up on the campus of Tongji University. The main purpose of this facility is to test the potential of various sensors in detecting early-warning signals.

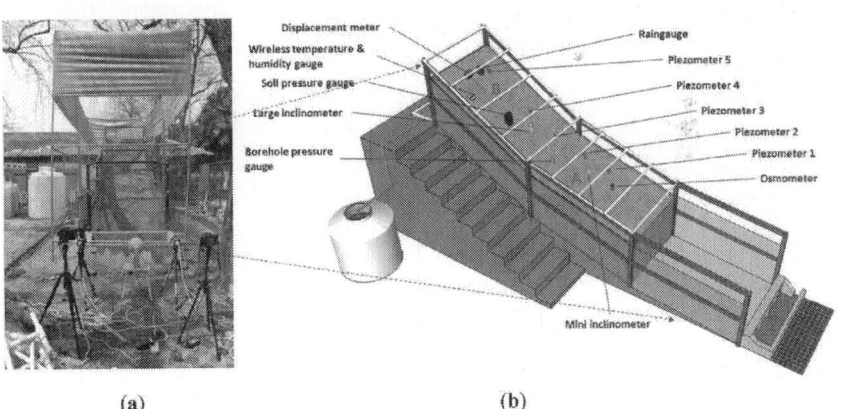

(a)   (b)

Figure 8. The scaled-down landslide simulation platform on the campus of Tongji University (**a**) and the spatial distribution of *in situ* sensors for the experiment (**b**).

This scaled-down landslide simulation experiment system is composed of a simulation platform of mass movement, a spatial SN, a data collection and communication sub-system, a data storage server, and a visualization system [33]. Firstly, the simulation platform was designed to simulate the movement type, the geological configurations, and the triggering factors of intensive rainfall for the real-scene landslide that occurred in Taziping, Sichuan Province. Secondly, the SN includes a set of contact and non-contact sensors that were installed for observing and measuring different types of parameters that are related to landslide initialization, reactivation, acceleration, and, finally, failure of the slope. Thirdly, the data collection and communication sub-system was composed of the following components: (1) a data taker (dataTaker®, DT80) to collect data from all the contact sensors and then wirelessly transfer the signals through a General Packet Radio Service (GPRS) network; (2) a serial-port sever to gather some of the non-contact sensors and then transmit to the database server through 3G service, and (3) a self-organizing and self-healing mesh network to collect and transfer the video stream in real time through an iMesh device developed by the OTEC Communication Technology Co. (Guangzhou, China) [61]. Fourthly, the data storage server receives data from the wireless network and then interprets and stores this data for different sensors into different categories of the database. Finally, a visualization system composed of nine screens (46 inches each) was installed in the monitoring headquarters. This can instantly visualize the sensor positions, real-time video stream of mass movement, 2D spatial distribution of piezometers, temporal readings of sensors and the status of different sensors.

To simulate the Taziping landslide, the platform dimensions were designed to be 1.5 m in width and 6 m in length. The platform was divided into three equal sections having slope angles of 5°, 15°, and 30°, respectively. Tempered glass windows framed in steel in both sides of the platform were fixed to hold the landslide mass, and for the visual observation of the internal changes. To simulate the rainfall-triggered landslide, an artificial precipitation system was installed at the top of the landslide body. This system is composed of a water tank of 3,000 L in volume for the water supply, a pump to lift water for rainfall, and five evenly distributed water sprays that can be controlled and manipulated for precipitation duration and intensity. In addition, a steel frame was constructed to support a water-tight canvas enclosure to keep the sprayed water within the landslide body.

In this experiment, a total of 13 spatial *in situ* sensors were deployed (Figure 8). These sensors were used mainly to monitor landslide characteristics and triggering factors such as surface displacements on the landslide body, slope inclination, pore water pressure of groundwater flow, rainfall intensity, soil pressure, acceleration of slope movement, *etc.* For example, five piezometers were aligned along the sliding direction at the bottom of the landslide body with a depth of 1.5 m so as to interpret the changes in pore water pressure before and after slope failure. However, due to the limitations of scale, some sensors that will be used in the field, such as the GNSS, are not meaningful in a scaled-down model, so these kinds of sensors were not tested for the experiments. In addition to *in situ* sensors, in order to observe the surface displacement and landslide failure process and to evaluate the landslide magnitude, some stereo-camera systems and 3D imaging sensors were integrated into the simulation platform. Technological aspects that should be tackled on the real-scene landslides require a more in-depth analysis before moving from the simulation platform to the field.

Two technological issues need to be better focused before illustrating the results from the simulation experiments. The first concerns the use of wired connections between each contact sensor and the corresponding data acquisition unit (chiefly the data Taker®), which seems to contradict the use of a Wireless Sensor Network (WSN). For real-time field applications, the use of clusters of sensors has been devised to be the optimal solution to coping with the following problems. First, a solar power supply (consisting of a solar panel and its control and storage unit) can be shared among a large number of sensors; based on this hypothesis, power cables would be required for energy supply, and thus can be used for data communication as well. A local acquisition unit can be located close to the power source and operate for the entire sensor cluster; the successive communication from the local control unit to a central control unit outside the field then can be based on wireless technology. From this point of view, the simulation platform reproduces a sensor cluster. Second, the deployment of several sensors in a neighborhood allows for the cross-validation of spatially correlated observations. The second technological concern is related to the use of oversized communication channels. This selection is motivated by the required scalability of the SN, which should be able to accommodate a variable number of sensors under the same network structure.

The significant number of sensor observations was derived from the rainfall induced-landslide simulation experiment conducted on 18 November 2011. This experiment started at 14:08 and lasted about 2 h and 40 min. Due to a wireless communication technical problem, there was a period of data loss lasting for 25 min (14:37–15:02). The experiment was performed under the simulation of an intensive rainfall of approximate 280 mm within 3 h (as recorded by the rain gauge), with two periods of extremely heavy rainfall (14:25–14:40, 15:40–15:50) (Figure 9).

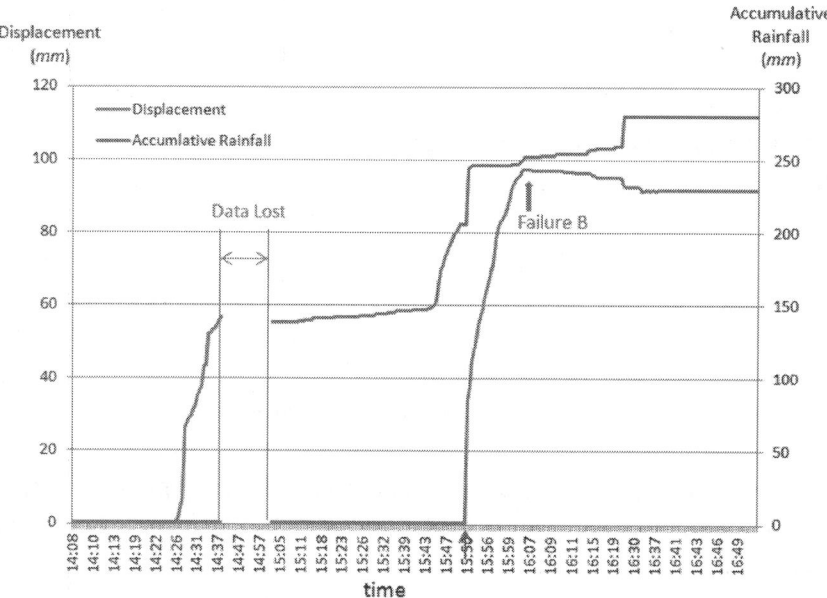

Figure 9. Monitoring data from the rain gauge and the displacement meter.

The flume test for a simulated landslide always renders the research concern on the rainfall condition, which necessarily needs a consideration of the field site and the amount of rainfall to trigger the slope failure in the experiment. As a result, it is necessary to set rainfall conditions that can evaluate the relationship between actual rainfall amount in the field and slope failure during the flume test. The regional area where the Taziping landslide is located is along the seismic zones that were strongly affected by the 2008 Wenchuan Earthquake. In this area, the amount of rainfall necessary to trigger landslides and debris flows nevertheless has not been well studied by previous researchers. [32] reported an accumulative rainfall of approx. 330 mm in 40 h (23–24 September 2008) which induced 969 new landslides in this seismic region. In our simulation, a rainfall accumulation of 280 mm over 3 h was used, with higher hourly rainfall intensity than the actual amounts in the field. The reason for using higher rainfall intensity during the experiment was to include the consideration of landslide scale. The simulated landslide is at a much smaller scale than the real-scene landslide. To match the scale, the rainfall simulation facility has been designed to produce decreased sizes of raindrops and increased density of raindrops. This is also the reason why natural rainfall is not used in the experiment considering its larger size of raindrop. In terms of the applied tipping bucket rain gauge with the same diameter (203 mm) of cylindrical container, it records higher rainfall intensity for dense, small-sized raindrops. This does not indicate that the measurements from the rain gauge are not correct, but it reflects the fact that in order to simulate the same effect of rainfall as in the field, a scaled rainfall simulation system with higher rainfall intensity is needed.

The first slope failure started from the bottom of the landslide body (Figure 8) at 15:50. This was observed by the borehole pressure gauge mounted at the bottom of the landslide with a depth of 1.5 m (Figure 10) and the mini inclinometer (Figure 11) set up in the lower part of the slope with a depth of 0.5 m. These gauges detected significantly increasing signals starting 5 min and 20 min, respectively, before the failure in section A. A subsequent slope failure occurred in the upper part of the slope at 16:05. This was recorded by the soil pressure gauge installed just below the slope surface (Figure 10), the displacement meter put on the slope surface (Figure 9), and the two-axis large inclinometer covering the whole depth of the landslide (Figure 11). These gauges noted a sudden signal change 25 min, 15 min and 20 min, respectively, before the failure in section B. In addition, the mass movement can be noticed 20 min before the initial failure by changes in the pore water pressure (Figure 12) that were measured by an osmometer and a set of piezometers affixed to the bottom the landslide body with a depth of 1.5 m.

Figure 10. Monitoring result of the borehole pressure gauge and the soil pressure gauge.

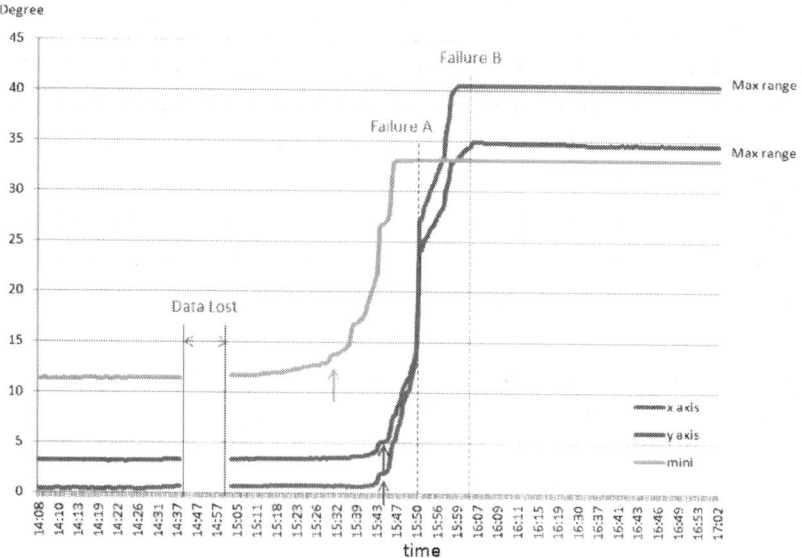

Figure 11. Monitoring data from the two-axis large and one-axis mini inclinometers.

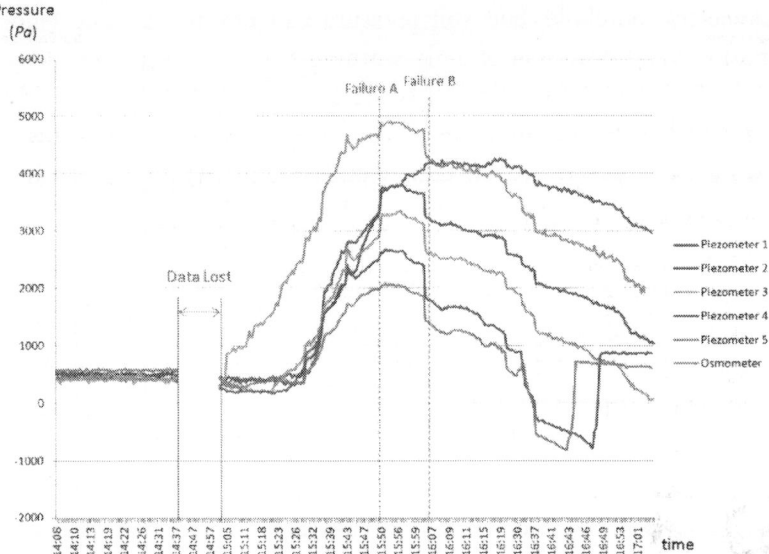

Figure 12. Monitoring data collected by the osmometer and the piezometers.

While the high-resolution cameras provide images throughout the entire duration of a simulation experiment, high-speed cameras are used mainly to analyze the final slope failure, which might feature very rapid movements. Both camera systems gather sequences of stereo images to be used for photogrammetric analysis (volume computation, surface point tracking); at the moment, these processing tasks are applied after the end of the experiment, in a manner similar to how satellite remote-sensing observations will be used in field applications. On the other hand, the high-speed camera system could be used for both tasks after selection of the most suitable time span of the images to process for each part of the experiment.

In recent years, several 3D imaging sensors have been introduced that can provide the direct acquisition of a 3D surface by means of phase-shift analysis (range cameras) or by using triangulation principles [62]. Although the resolution and precision are lower than the ones obtained from the above-mentioned photogrammetric systems, they would allow for real-time processing.

In a nutshell, the above-mentioned landslide simulation experiment confirms the usefulness of *in situ* sensors (such as inclinometers, piezometers, borehole, and soil pressure gauges) in real-time landslide monitoring and detection of early-warning signals. These sensors can be further employed reliably in real-scene landslide monitoring in Taziping, Sichuan Province, China. In addition, the experiment shows the possibility of transferring real-time video streams from the infrared video camera through a wireless network. As new approaches for landslide monitoring, high-speed stereo cameras and 3D imaging sensors can provide alternative means for landslide stability evaluation through remote-sensing approaches. However, due to scale differences and field conditions, they need to be further tested and evaluated in the field for real-scene landslide monitoring.

IN SITU SENSOR NETWORK DEPLOYMENT IN TAZIPING LANDSLIDE

The Existing Landslide Prevention Facilities and Monitoring Sensors

The Taziping landslide (Figure 13) is located above Hongkou Town, which was strongly affected by the 2008 Wenchuan Earthquake. The elevations of the source and the toe areas are about 1,370 m and 1,007 m, respectively, with a relative elevation difference of about 363 m. The slope ranges from 25° to 40°, with relatively steep landslide source areas and a relatively flat toe of mass accumulation. After the event, the landslide was frequently accelerated by intensive rainfall and subsequent water infiltrations, especially with the accumulation of loose debris and the fragile soil characteristics triggered by the earthquake [32]. Due to the extensive amounts of loose debris carried from the source areas, this landslide has the potential to turn into debris flows with intensive surface water adjunction. In addition, frequent human activities such as road building, forest cutting and fruit cultivation can often further decrease the stability of the slope and accordingly accelerate the sliding processes. The Taziping landslide represents a major threat to the town at the foot of the mountain not only in terms of human lives, but also in terms of infrastructure and local economic development.

	Mitigation Wall		Drainage Trench
	Landslide Boundary	▲	Soil Moisture Sensors
●	Surface Clinometers	●	Inclinometer

Figure 13. An overview of the Taziping landslide, the mitigation infrastructure and the currently available sensors for monitoring.

Due to the high vulnerability of the slope, since 2010 the Taziping landslide has been continuously reinforced by the constructions of mitigation infrastructure. Efforts include the positioning of anchors, mitigation walls, shallow drainage trenches and a system for surface water run-off (Figures 13 and 14). In order to monitor the effectiveness of these mitigation efforts, three types of preliminary sensors have been installed on the landslide. First, a MEMS (Micro-Electro-Mechanical Systems) inclinometer with a depth of 15 m has been set up in the middle of the slope to measure the sliding surface and evaluate the activeness of the landslide. Second, three surface clinometers were established close to the mitigation wall in order to measure the surface displacement of the slope and evaluate the effectiveness of the mitigation program. Third, for the purpose of measuring ground water, three soil moisture sensors were mounted close to the surface drainage system. Although these sensors provide a useful approach to monitor the status of the slope, the system does not provide the possibility for real-time transfer of acquired data and it requires frequently manual readings of data collected by *in situ* data collection units. As a result, a more sophisticated real-time monitoring system is needed.

Figure 14. Some existing facilities and sensors at the Taziping landslide: (**a**) Installation for soil moisture sensor; (**b**) Inclinometer; (**c**) Mitigation wall; and (**d**) Drainage trench.

Design Considerations for Future Sensor Network Deployment at the Taziping Landslide

The SN tested in the landslide simulation experiment was successful in the following aspects. First, it enabled a rapid data transferring from the landslide to the monitoring headquarters in real time, which satisfied emergency management requirements when the slope is approaching failure. Second, a combination of three communication approaches (GPRS, 3G and broadband) separated different types of data and thus avoided transfer congestion at a single data node. Third, the experiment showed that a network-based monitoring system is more reliable than a point-based strategy because sensors can work collaboratively and cross-validation of data reliability is made possible. Fourth, the networked sensors ensure the transmission efficiency and accordingly avoid great delay of data transferring and storage. Finally, the experiment encompasses a complete and sophisticated monitoring system that includes *in situ*SN, data collection units, diverse communication approaches, data processing and storage units, and a visualization system.

Such a system will be transferred in the future to real-scene landslide monitoring in Taziping. A preliminary design of the improved network is sketched in Figure 15. The plan of this network is to use five communication relay stations on the slope of the landslide, each of which serves as a gateway node, namely working as a local data collection unit and as an intermediate point for data distribution and communication. These five stations are designed to form a connected SN, provided that the data communication between them can be maintained. If there are significant obstacles (e.g., forests) which block communication routes, more stations will be established, with each installation maintaining a certain distance above the ground. Each station will apply a multi-hop device for fast roaming communication that allows for a fast shift (within 0.01 s).

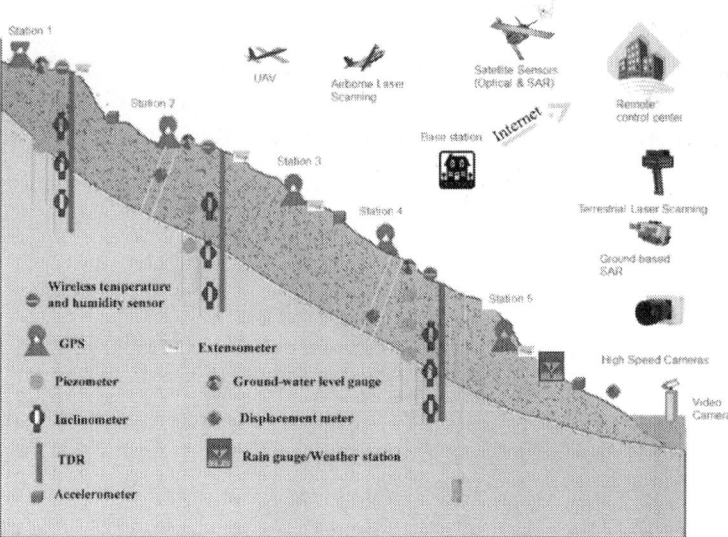

Figure 15. Design of the spatial sensor network to be deployed at the Taziping landslide.

For each sensor node of the station, a series of sensors will be connected for local data storage and distribution. Inside the drilled borehole, piezometers will be installed to monitor changes of pore water pressure in the ground water, especially for real-time monitoring of water infiltration after an intensive rainfall event. Also, inclinometers will be assembled and then installed inside the slope so as to measure slope movement both vertically and horizontally. Each gateway node also will include an extensometer that measures surface displacement. In terms of a rainfall triggering factor, a rain gauge will be used to monitor accumulative rainfall in real time. By employing a weather station that takes advantage of the National Oceanic and Atmospheric Administration (NOAA) weather forecast model [63], it is possible to forecast rainfall levels within 12 h, thus allocating the monitoring priority to the rain gauge and the weather station. These sensors were well tested during the landslide simulation experiment and can reach expected results for landslide monitoring.

In addition, the following sensors have been considered for more comprehensive observations. Time Domain Reflectometry (TDR) is to be installed to detect internal fractures of the slope. In the meanwhile, it can measure the level of ground water, providing comparisons for the observations from the piezometers. In addition, plans call for mounting high-frequency (80 Hz) accelerometers to be used for detecting surface vibration caused by the mass movement and, subsequently, for evaluation of the amount of acceleration. If the satellite signals can be received, GNSS then can be added as a ground reference for measuring surface displacement. Besides *in situ* sensors, certain remote-sensing platforms are scheduled to be installed. A ground-based InSAR [64,65] will be installed at the toe of the landslide so as to scan the entire slope and measure surface displacement along the Line-Of-Sight (LOS). Moreover, a pair of stereo high-speed cameras has already been tested and will be employed to capture the moment of slope failure so as to further analyze the landslide mechanism especially during the critical moment of slope failure. Since topographic information concerning a slope is fundamental for further analysis of landslide process, a long-range terrestrial laser scanning system [66] is to be put at the foot of the mountain so as to obtain a high-resolution DTM of the slope and, with multi-temporal acquisitions, to measure the surface displacement from both cloud points and sequential DTMs. The point clouds can also be acquired from airborne laser scanning [67], which allows for a larger scanning extent useful for regional analysis. Owing to its flexibility and low cost, it is planned to employ a Unmanned Aerial Vehicle (UAV) [68], embedded with a high-resolution camera, to acquire high-resolution aerial photos, to detect small rock fissures of the slope, and to generate a DTM for comparison with that formed from laser scanning. Finally, based on the satellite platforms, the ground movement can also be well detected using InSAR [69], especially for shorter wavelength sensors such as TerraSAR-X and COSMO-SkyMed, which allows a detection precision of millimeters.

Nevertheless, for a successful transplantation of the SN from the simulation platform to the *in situ* environment, careful consideration still needs to be made, particularly in the following aspects. First, the power supply system using solar panels needs to be established and it is necessary to ensure all sensor and gateway nodes can be supplied with more than 20% residuals during rainy or cloudy days. Second, for remote control of the installed sensors, an XML file is needed to have sending and editing functions, aiming at an adjustment of sensor parameters through the client server. Third, in case of data lost, it is crucial to establish a VPN transferring route between the local data center in Taziping and the monitoring headquarters in Shanghai, to ensure the safety of data security and data transfer. Fourth, a synchronization approach using GNSS should be tested so as to ensure that the synchronization rate is within 0.1 ms. Finally, when there is network congestion, the priority of data transfer needs to be defined with a first consideration of critical sensors relating to monitoring of the mass movement.

CONCLUSIONS

In this study, a systematic approach to landslide investigation based on remote sensing and Sensor Network (SN) has been developed and assessed. This approach is composed of three parts, landslide susceptibility mapping for susceptible spots determination in Hongkou Town (Sichuan, China), scaled-down landslide simulation for SN prototype system test on the campus of Tongji University (Shanghai,

China), and *in situ* SN design for field landslide observation at the Taziping landslide, Hongkou Town. First, the landslides triggered by the 2008 Wenchuan Earthquake were extracted by means of an object-oriented method from a pair of geo-referenced pre- and post- event IKONOS images (3.25 km² of landslide area compared with the 30 km² study site), followed by the correlation analysis between the landslide distribution and the control factors generated by DEM and vector data. The landslide susceptibility map was then produced by the Artificial Neural Network (ANN) algorithm to determine the susceptible landslide spots, and the Taziping landslide was selected for further investigation and SN deployment. Second, the SN was tested on a scaled-down landslide simulation platform before the *in situ*deployment to examine the usability of the sensors as well as the deployment strategy. Finally, the considerations and suggestions were given for the future *in situ* SN deployment in Taziping.

The research results support the following conclusions:

1. The spatial distribution of earthquake-triggered landslides is found to correlate with some of the control factors, e.g., it has positive correlation with the slope and seismic intensity factors, and the landslides occurred more easily in valleys with flowing rivers as well as within 200-to-500 m distance to roads.

2. The ANN is proved to be effective for generation of landslide susceptibility map to identify the zones of high risk. Correspondences of estimated susceptible areas compared with landslides extracted from the high-resolution IKONOS images were found in 89% of the cases.

3. The landslide simulation experiment demonstrates the applicability of the *in situ* sensors and SN in real-time landslide monitoring triggered by rainfall, and the potential for detection of early-warning signals useful for landslide prediction before the slope failure (e.g., 10 to 20 min in advance).

4. The integration of remote sensing approach and SN technique contributes to landslide research by bridging the gaps between large-scale regional landslide investigation and small-scale individual landslide monitoring, allowing the localization of slopes with higher

hazard-risk probability for intensified and real-time landslide observation with deployment of SN.

Despite the achievements in this research, there are currently several limitations that need further improvement in the future. For example, the relatively short pre-alerting time requires the use of real-time data collection, communication and analysis to be exploited in the monitoring system for the deployment of SN in Taziping. Due to scale differences and field conditions between the simulation platform and the slope in Taziping, further tests in the field are required for real-scene landslide monitoring. Finally, the proposed methodology suffers from approximations entailed in the scaled-down models. These approximations, which however would be involved in any other modeling techniques (e.g., in numerical or empirical modeling), call for a careful validation with respect to observations collected from the real landslide in the future.

This study is supported by the National High Technology Research and Development Program of China (2012AA121302), the State Key Development Program for Basic Research of China (2012CB957701, 2013CB733204 and 2013CB733203), the National Science Foundation of China (41201425, 41201424, and 41101443), the Key Laboratory of Advanced Engineering Surveying of SBSM (TJES0905) and the Key Laboratory of Mapping from Space of SBSM (K201005), and the Key Laboratory for Land Environment and Disaster Monitoring of SBSM (LEDM2009B02). The authors would like to thank Leslie B. Smith from the Mapping and GIS Lab, OSU for her English editing of the manuscript, and the anonymous reviewers for their valuable comments and suggestions.

REFERENCES

1. Schuster, R.L. Socioeconomic Significance of Landslides. In *Landslides: Investigation and Mitigation*; Turner, A.K., Schuster, R.L., Eds.; Transportation Research Board Special Report 247;. National Academies Press: Washington, DC, USA, 1996; pp. 12–35.

2. Sassa, K.; Tsuchiya, S.; Ugai, K.; Wakai, A.; Uchimura, T. Landslides: A review of achievements in the first 5 years (2004–2009). *Landslides* **2009**, *6*, 275–286.

3. Delacourt, C.; Allemand, P.; Berthier, E.; Raucoules, D.; Casson, B.; Grandjean, P.; Pambrun, C.; Varel, E. Remote-sensing techniques for analysing landslide kinematics: A review. *Bull. Soc. Géol. Fr* **2007**, *178*, 89–100.

4. Metternicht, G.; Hurni, L.; Gogu, R. Remote sensing of landslides: An analysis of the potential contribution to geo-spatial systems for hazard assessment in mountainous environments. *Remote Sens. Environ* **2005**, *98*, 284–303.

5. Guzzetti, F.; Mondini, A.C.; Cardinali, M.; Fiorucci, F.; Santangelo, M.; Chang, K.-T. Landslide inventory maps: New tools for an old problem. *Earth Sci. Rev* **2012**, *112*, 42–66.

6. Tung, S.-H.; Shih, M.-H.; Sung, W.-P. Identification of the landslide using the satellite images and the digital image correlation method. *Disa. Adv* **2013**, *6*, 4–9.

7. Van Westen, C.J.; Castellanos, E.; Kuriakose, S.K. Spatial data for landslide susceptibility, hazard, and vulnerability assessment: An overview. *Eng. Geol* **2008**, *102*, 112–131.

8. Felicísimo, M.Á.; Cuartero, A.; Remondo, J.; Quirós, E. Mapping landslide susceptibility with logistic regression, multiple adaptive regression splines, classification and regression trees, and maximum entropy methods: A comparative study. *Landslides* **2013**, *10*, 175–189.

9. Strozzi, T.; Ambrosi, C.; Raetzo, H. Interpretation of aerial photographs and satellite SAR interferometry for the inventory of landslides. *Remote Sens* **2013**, *5*, 2554–2570.

10. Chauhan, S.; Sharma, M.; Arora, K.M. Landslide susceptibility zonation of the Chamoli region, Garhwal Himalayas, using logistic regression model. *Landslides* **2010**, *7*, 411–423.

11. Park, S.; Chio, C.; Kim, B.; Kim, J. Landslide susceptibility mapping using frequency ratio, analytic hierarchy process, logistic regression, and artificial neural network methods at the Inje area, Korea. *Environ. Earth Sci* **2013**, *68*, 1443–1464.

12. Tofani, V.; Raspini, F.; Catani, F.; Casagli, N. Persistent Scatterer Interferometry (PSI) technique for landslide characterization and monitoring. *Remote Sens* **2013**, *5*, 1045–1065.

13. Mantovani, F.; Soeters, R.; van Westen, C.J. Remote sensing techniques for landslide studies and hazard zonation in Europe. *Geomorphology* **1996**, *15*, 213–225.

14. Dunning, S.A.; Rosser, N.J.; Massey, C.I. The integration of terrestrial laser scanning and numerical modeling in landslide investigations. *Q. J. Eng. Geol. Hydrogeol* **2010**, *43*, 233–247.

15. Gigli, G.; Fanti, R.; Canuti, P.; Casagli, N. Integration of advanced monitoring and numerical modeling techniques for the complete risk scenario analysis of rockslides: The case of Mt. Beni (Florence, Italy). *Eng. Geol* **2011**, *120*, 48–59.

16. Li, R.; The CSISSD Research Team. Advanced Spatial Sensor Network Systems—Review, Status, and Applications. Proceedings of 2011 International Symposium on Image and Data Fusion, Tengchong, Yunnan, China, 9–11 August 2011.

17. Ramesh, M.V. Design, development, and deployment of a wireless sensor network for detection of landslides. *Ad Hoc Netw.* **2012**.

18. Arattano, M.; Marchi, L. Systems and sensors for debris-flow monitoring and warning. *Sensors* **2008**, *8*, 2436–2452.

19. Angeli, M.; Pasuto, A.; Silvano, S. A critical review of landslide monitoring experiences. *Eng. Geol* **2000**, *55*, 133–147.

20. Akbarimehr, M.; Motagh, M.; Haghshenas-Haghighi, M. Slope stability assessment of the Sarcheshmeh Landslide, Northeast Iran, investigated using InSAR and GPS observations. *Remote Sens* **2013**, *5*, 3681–3700.

21. Jaboyedoff, M.; Oppikofer, T.; Abellán, A.; Derron, M.H.; Loye, A.; Metzger, R.; Pedrazzini, A. Use of LIDAR in landslide investigations: A review. *Nat. Hazards* **2012**, *61*, 1–24.

22. Lato, M.J.; Bevan, G.; Fergusson, M. Gigapixel imaging and photogrammetry: Development of a new long range remote imaging technique. *Remote Sens* **2012**, *4*, 3006–3021.

23. del Ventisette, C.; Ciampalini, A.; Manunta, M.; Calò, F.; Paglia, L.; Ardizzone, F.; Mondini, A.C.; Reichenbach, P.; Mateos, R.M.; Bianchini, S.; *et al.* Exploitation of large archives of ERS and ENVISAT C-Band SAR data to characterize ground deformations. *Remote Sens* **2013**, *5*, 3896–3917.

24. Ou, J.; Qiao, G.; Bao, F.; Wang, W.; Di, K.; Li, R. A new method for automatic large scale map updating using mobile mapping imagery. *Photogramm. Rec* **2013**, *28*, 240–260.

25. Sato, H.P.; Harp, E.L. Interpretation of earthquake-induced landslides triggered by the 12 May 2008, M7.9 Wenchuan earthquake in the Beichuan area, Sichuan Province, China, using satellite imagery and Google Earth. *Landslides* **2009**, *6*, 153–159.

26. Bozzano, F.; Cipriani, I.; Mazzanti, P.; Prestininzi, A. Displacement patterns of a landslide affected by human activities: Insights from ground-based InSAR monitoring. *Nat. Hazards* **2011**, *59*, 1377–1396.

27. Hong, Y.; Adler, R.F. Towards an early-warning system for global landslides triggered by rainfall and earthquake. *Int. J. Remote. Sens* **2007**, *28*, 3713–3719.

28. Cruden, D.M.; Varnes, D.J. Landslides Types and Processes. In *Landslides: Investigation and Mitigation*; Turner, A.K., Schuster, R.L., Eds.; Transportation Research Board Special Report 247,. National Academies Press: Washington, DC, USA, 1996; pp. 36–75.

29. Guzzetti, F.; Carrara, A.; Cardinali, M.; Reichenbach, P. Landslide hazard evaluation: A review of current techniques and their application in a multi-scale study, Central Italy. *Geomorphology* **1999**, *31*, 181–216.

30. Dai, F.C.; Lee, C.F.; Nhai, Y.Y. Landslide risk assessment and management: An overview. *Eng. Geol* **2002**, *64*, 65–87.

31. Dai, F.C.; Xu, C.; Yao, X.; Xu, L.; Tu, X.B.; Gong, Q.M. Spatial distribution of landslides triggered by the 2008 Ms 8.0 Wenchuan Earthquake, China. *J. Asian Earth Sci* **2011**, *40*, 883–895.

32. Tang, C.; Zhu, J.; Qi, X.; Ding, J. Landslides induced by the Wenchuan earthquake and the subsequent strong rainfall event: A case study in the Beichuan area of China. *Eng. Geol* **2011**, *122*, 22–33.

33. Scaioni, M.; Lu, P.; Chen, W.; Qiao, G.; Wu, H.; Feng, T.; Tong, X.; Wang, W.; Li, R. Analysis of spatial sensor network observations during landslide simulation experiments. *Eur. J. Environ. Civil. Eng.* **2013**.

34. Qi, S.W.; Xu, Q.; Lan, H.X.; Zhang, B.; Liu, J.Y. Spatial distribution analysis of landslides triggered by 2008.5.12 Wenchuan Earthquake, China. *Eng. Geol* **2010**, *116*, 95–108.

35. Wang, F.W.; Cheng, Q.G.; Highland, L.; Miyajima, M.; Wang, H.B.; Yan, C.G. Preliminary investigation of some large landslides triggered by the 2008 Wenchuan earthquake, Sichuan Province, China. *Landslides* **2009**, *6*, 47–54.

36. Qiao, G.; Wang, W.; Wu, B.; Liu, C.; Li, R. Assessment of geo-positioning capability of high-resolution satellite imagery for densely populated high buildings in metropolitan areas. *Photogramm. Eng. Remote. Sens* **2010**, *76*, 923–934.

37. Lu, P.; Stumpf, A.; Kerle, N.; Casagli, N. Object-oriented change detection for landslide rapid mapping. *IEEE Geosci. Remote Sens* **2011**, *8*, 701–705.

38. Hölbling, D.; Füreder, P.; Antolini, F.; Cigna, F.; Casagli, N.; Lang, S. A semi-automated object-based approach for landslide detection validated by Persistent Scatterer Interferometry measures and landslide inventories. *Remote Sens* **2012**, *4*, 1310–1336.

39. Yang, C.J.L.; Ren, X.L.; Huang, H. The vegetation damage assessment of the Wenchuan earthquake of May 2008 using remote sensing and GIS. *Nat. Hazards* **2012**, *62*, 45–55.

40. Lacroix, P.; Zavala, B.; Berthier, E.; Audin, L. Supervised method of landslide inventory using panchromatic SPOT5 images and application to the earthquake-triggered landslides of Pisco (Peru, 2007, Mw8.0). *Remote Sens* **2013**, *5*, 2590–2616.

41. Defries, S.R.; Townshend, J.R.G. NDVI-derived land cover classifications at a global scale. *Int. J. Remote Sens* **1994**, *15*, 3567–3586.

42. Marqués, F.; Gasull, A. *Multiresolution Image Segmentation Based on Compound Random Fields: Application to Image Coding*; Universitat Politècnica de Catalunya: Barcelona, Spain, 1992; pp. 86–114.

43. Tong, H.J.; Maxwell, T.; Zhang, Y.; Dey, V. A supervised and fuzzy-based approach to determine optimal multi-resolution image segmentation parameters. *Photogramm. Eng. Remote. Sens* **2012**,*78*, 1029–1044.

44. Gorum, T.; Fan, X.M.; van Westen, J.C.; Huang, R.Q.; Xu, Q.; Tang, C.; Wang, G.H. Distribution pattern of earthquake-induced landslides triggered by the 12 May 2008 Wenchuan earthquake.*Geomorphology* **2011**, *133*, 152–167.

45. Conoscenti, C.; Maggio, C.D.; Rotigliano, E. GIS analysis to assess landslide susceptibility in a fluvial basin of NW Sicily (Italy). *Geomorphology* **2008**, *94*, 325–339.

46. Neuhäuser, B.; Damm, B.; Terhorst, B. GIS-based assessment of landslide susceptibility on the base of the Weights-of-Evidence model. *Landslides* **2012**, *9*, 511–528.

47. Pareek, N.; Sharma, M.L.; Arora, K.M. Impact of seismic factors on landslide susceptibility zonation: A case study in part of Indian Himalayas. *Landslides* **2010**, *7*, 191–201.

48. Chen, H.; Lin, G.W.; Lu, M.H.; Shih, T.Y.; Horng, M.J.; Wu, S.J.; Chuang, B. Effects of topography, lithology, rainfall and earthquake on landslide and sediment discharge in mountain catchments of southeastern Taiwan. *Geomorphology* **2011**, *133*, 132–142.

49. Akgun, A. A comparison of landslide susceptibility maps produced by logistic regression, multi-criteria decision, and likelihood ratio methods: A case study at İzmir, Turkey. *Landslides* **2012**, *9*, 93–106.

50. Othman, A.A.; Gloaguen, R. River courses affected by landslides and implications for hazard assessment: A high resolution remote sensing case study in NE Iraq–W Iran. *Remote Sens* **2013**, *5*, 1024–1044.

51. Moreiras, S.M. Landslide susceptibility zonation in the Rio Mendoza Valley, Argentina.*Geomorphology* **2005**, *66*, 345–357.

52. Akgun, A.; Dag, S.; Bulut, F. Landslide susceptibility mapping for a landslide-prone area (Findikli, NE of Turkey) by likelihood-frequency ratio and weighted linear combination models.*Environ. Geol* **2008**, *54*, 1127–1143.

53. Ray, R.L.; Jacobs, J.M.; Cosh, M.H. Landslide susceptibility mapping using downscaled AMSR-E soil moisture: A case study from Cleveland Corral, California, US. *Remote Sens. Environ* **2010**,*114*, 2624–2636.

54. Bartlett, E.B. A stochastic training algorithm for artificial neural networks. *Neurocomputing* **1994**,*6*, 31–43.

55. Li, Y.; Chen, G.; Tang, C.; Zhou, G.; Zheng, L. Rainfall and earthquake-induced landslide susceptibility assessment using GIS and Artificial Neural Network. *Nat. Hazard Earth Syst. Sci***2012**, *12*, 2719–2729.

56. Liu, C.; Li, W.; Wu, H.; Lu, P.; Sang, K.; Sun, W.; Chen, W.; Hong, Y.; Li, R. Susceptibility evaluation and mapping of China's landslides based on multi-source data. *Nat. Hazards* **2013**.

57. Holmstrom, L.; Koistinen, P. Using additive noise in back-propagation training. *IEEE Trans. Neural Netw* **1992**, *3*, 24–38.

58. Haykin, S. *Neural Networks: A Comprehensive Foundation*, 2nd ed; Prentice-Hall: Upper Saddle River, NJ, USA, 1999; p. 823.

59. Ma, C.F.; Jiang, L.H. Some research on Levenberg–Marquardt method for the nonlinear equations.*Appl. Math. Comput* **2007**, *184*, 1032–1040.

60. Lee, S.; Ryu, J.H.; Kim, L.S. Landslide susceptibility analysis and its verification using likelihood ratio, logistic regression, and artificial neural network models: Case study of Youngin, Korea.*Landslides* **2007**, *4*, 327–338.

61. OTEC Communication Technology Co. (Guangzhou, China). Available online http://www.cnotec.cn/1234/html/?18.html (assessed on 3 September 2013)..

62. Sansoni, G.; Trebeschi, M.; Docchio, F. State-of-the-art and applications of 3D imaging sensors in industry, cultural heritage, medicine, and criminal investigation. *Sensors* **2009**, *9*, 568–601.

63. NOAA National Weather Service. Available online http://mag.ncep.noaa.gov/ (accessed on 3 September 2013)..

64. Casagli, N.; Catani, F.; del Ventisette, C.; Luzi, G. Monitoring, prediction, and early warning using ground-based radar interferometry. *Landslides* **2010**, *7*, 291–301.

65. Tapete, D.; Casagli, N.; Luzi, G.; Fanti, R.; Gigli, G.; Leva, D. Integrating radar and laser-based remote sensing techniques for monitoring structural deformation of archaeological monuments. *J. Archaeol. Sci* **2013**, *40*, 176–189.

66. Heritage, G.L.; Large, A.R.G. *Laser Scanning for the Environmental Sciences*; John Wiley & Sons: Chichester, UK, 2009; p. 302.

67. Vosselman, G.; Maas, H.G. *Airborne and Terrestrial Laser Scanning*; Taylor and Francis Group: Boca Raton, FL, USA, 2010; p. 320.

68. Eisenbeiss, H.; Sauerbier, M. Investigation of UAV systems and flight modes for photogrammetric applications. *Photogramm. Rec* **2011**, *26*, 400–421.
69. Tapete, D.; Fanti1, R.; Cecchi, R.; Petrangeli, P.; Casagli, N. Satellite radar interferometry for monitoring and early-stage warning of structural instability in archaeological sites. *J. Geophys. Eng* **2012**, *9*, S10–S25.

CITATION

Gang Qiao , Ping Lu , Marco Scaioni, Shuying Xu, Xiaohua Tong, Tiantian Feng, Hangbin Wu, Wen Chen, Yixiang Tian, Weian Wang and Rongxing Li, Landslide Investigation with Remote Sensing and Sensor Network: From Susceptibility Mapping and Scaled-down Simulation towards in situ Sensor Network Design, doi:10.3390/rs5094319

CHAPTER 9

Multivariate Differencing Techniques for Land Cover Change Detection: the Normalized Difference Reflectance Approach

Paolo Villa[1], Giovanmaria Lechi[2] and Mario A. Gomarasca[1]

[1]Institute for Electromagnetic Sensing of the Environment (CNR-IREA), National Research Council, Milan,, Italy
[2]Bulding Environment Sciences and Technology (BEST) Department, Polytechnic of Milan, Italy

INTRODUCTION

The importance of the dynamic side of natural and man-made phenomena has become an urgent need when trying to mitigate the human impact on environment. Remote Sensing is one of the most effective way to quantify and map the changes of environmental conditions on our planet: the tools used for this purpose are called Change Detection Techniques. Techniques among which an important role is played by those methodologies based on multi-spectral remote sensing data and exploiting multivariate analysis derived methodologies, also demonstrating their capabilities through some test cases, covering flood events and urban growth studies.

Multi-temporal and multi-spectral techniques for Change Detection exist in a wide variety of approaches, often far too sector oriented and not straightforward. Compression and decorrelation techniques, on the other side, tend not to exploit the whole spectral content of remotely sensed data. The Normalized Difference Reflectance (NDR) here introduced is a general approach for bi-temporal land cover change mapping and detection that exploits the whole spectral capabilities of panchromatic, multi-spectral or hyper-spectral images. NDR is a general and simple measure that can be used in the frame of what are called Normalized Difference Change Detection Techniques (NDCD), which starts using as a input the NDR derived results. This Chapter includes a large test case which is a good benchmark for NDR approach, using Minimum Noise Fraction implementation of NDCD for mapping Hurricane Katrina aftermaths over the city of New Orleans, U.S., thus fusing together urban and flood change applications.

The purpose of the chapter is to give an overview of multivariate difference-based techniques for land cover change mapping using multispectral remote sensing data, and to introduce and demonstrate the Normalized Difference Reflectance approach in the frame of Normalized Difference Change Detection techniques. Two examples of NDCD results are given as a complement to theoretical aspects of the methodology, and an application study has been used as benchmark for the technique performances evaluation, in comparison with other established Change Detection techniques.

MULTIVARIATE DIFFERENCES IN CHANGE DETECTION

Multi-temporal differencing is an established change detection technique for environmental mapping and monitoring with remotely sensed data (Singh, 1989, Lu et al., 2004, Coppin et al., 2004). Following a difference normalization approach, introduced in remote sensing for vegetation studies with the normalized difference vegetation index (NDVI), a multi-temporal implementation of this standardization technique for forest change analysis was first proposed for univariate vegetation indexes (VIs) (Coppin & Bauer, 1994), and then in comparison with other change detection methodologies (Coppin et al., 2001), always for forest mapping purposes.

During this work it has been introduced a quantitative method to evaluate land cover change through multi-spectral variation in radiometric response of surface features. In order to detect interesting changes, a pair of satellite scenes, geometrically registered and atmospherically corrected, is to be radiometrically normalized. After that, a map of spectral variations is produced using a multi-spectral difference index named Normalized

Difference Reflectance (NDR). The NDR is therefore an approach to Change Feature Identification phase in Change Detection.

The phase of Change Mapping is then performed using NDR measures as inputs for Change Detection methodologies and techniques, in the frame of what are called Normalized Difference Change Detection (NDCD) techniques. The NDCD is a technique which, given an image pair, performs calculations on radiometric normalized reflectance data through the definition of the normalized difference reflectance (NDR) and produces a standardized difference of the reflectance values.

The use of NDR and NDCD will be presented through case studies showing change analysis covering flood events and urban environment: one case is over a flood event occurred in Bangladesh and exploiting Landsat-7/ETM+ scenes, another case regards the urban expansion scenario of Washington outskirts, U.S., and exploiting Terra ASTER data, the last and most complete case study is the analysis of the damages to the urban area of the city of New Orleans (Louisiana, USA) resulting from the passage of hurricane Katrina, using both SPOT-4/HRVIR and Landsat-5/TM data.

NORMALIZED DIFFERENCE REFLECTANCE (NDR)

The Normalized Difference Reflectance (NDR) here introduced is a general approach for bi-temporal land cover change mapping and detection that exploits the whole spectral capabilities of panchromatic, multi-spectral or hyper-spectral images. Given an image pair, the NDR produces a standardized difference by analyzing the changes in the reflectance properties of each spectral band, so without losing any spectral richness as when applying indexes, feature reduction or compression techniques (Villa & Lechi, 2007).

The image reflectance differences were modified to a normalized version on the sum of spectral values, in order to minimize the confusion among difference values which are numerically equal, but come from different land cover change events.

Hence, for every spectral band the NDR is defined as follows:

$$NDR_j = R_{jnorm}(post) - R_{jnorm}(pre)R_{jnorm}(post) + R_{jnorm}(pre)$$

(1)

where:

NDR = normalized difference reflectance

R^{norm} (post) = normalized reflectance for the post flood scene
R^{norm} (pre) = normalized reflectance for the pre flood scene
j = spectral band number

The NDR is a multi-spectral quantity which spans over the range of values from -1 to +1 and shows the amount of change in surface reflectance for every band in the original data, in terms of the relative difference in spectral signature of ground objects (-1.00 = maximum reflectance decrease, 0.00 = no change, +1.00= maximum reflectance increase). This approach is a quantitative base for building the successive phase of change detection, through the use of multi-spectral normalized reflectance values. A first and simple visual inspection of NDR band compositions permits a prompt and clear preliminary assessment of changes, thus supporting the choice of an apt change detection algorithm or technique for the phase of change mapping. Figure 1 shows an example of NDR calculated for calculated for an urban scenario, located in Maryland, U.S, using multi-temporal Terra ASTER data.

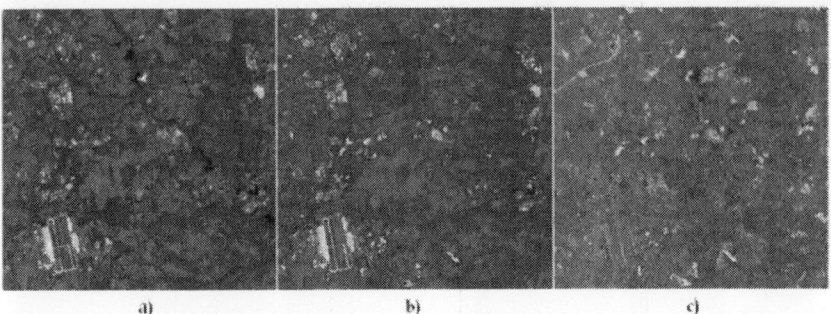

Figure 1. NDR calculated for an urban scenario, located in Maryland, U.S.: CIR visualization (RGB=3N,2,1), ASTER scene of April 9th, 2000 (a); CIR visualization (RGB=3N,2,1), ASTER scene of August 24th, 2003 (b); CIR visualization (RGB=3N,2,1), NDR values derived (c). Different colours in (c) are inked to different kind of variations in surface reflectance between (a) and (b) images: grey areas represent not changed features, cyan areas represent a decreasing response in near infrared (linked to newly exposed areas, construction sites and new impervious surfaces), red areas represent increasing near infrared response (linked to phenological conditions of vegetation, going to an April scene to an August one), white areas represent increased response in all the visualized bands.

This approach allow to promptly visualize in RGB channels different triplets of bands at a time, thus bringing the user to a straightforward inspection of multi-spectral change features of surface objects; beginning with this visualization of multi-date information every end-user has the possibility to decide which may be the best Change Detection Technique to retrieve a land cover change map. NDR not only permits an easy and straightforward multi-spectral comparison and evaluation of land cover changes, but permits enhanced individualization of radiometric response change, in comparison with simple Reflectance Differencing (RD). In fact, the same amount of reflectance difference between two surface features can be due to different land cover changes depending on the reference amount of spectral response. The NDR approach takes into account this issue and outputs different values of NDR for the same RD situation, when corresponding to different changes, as illustrated in the example of Figure 2 and Table 1.

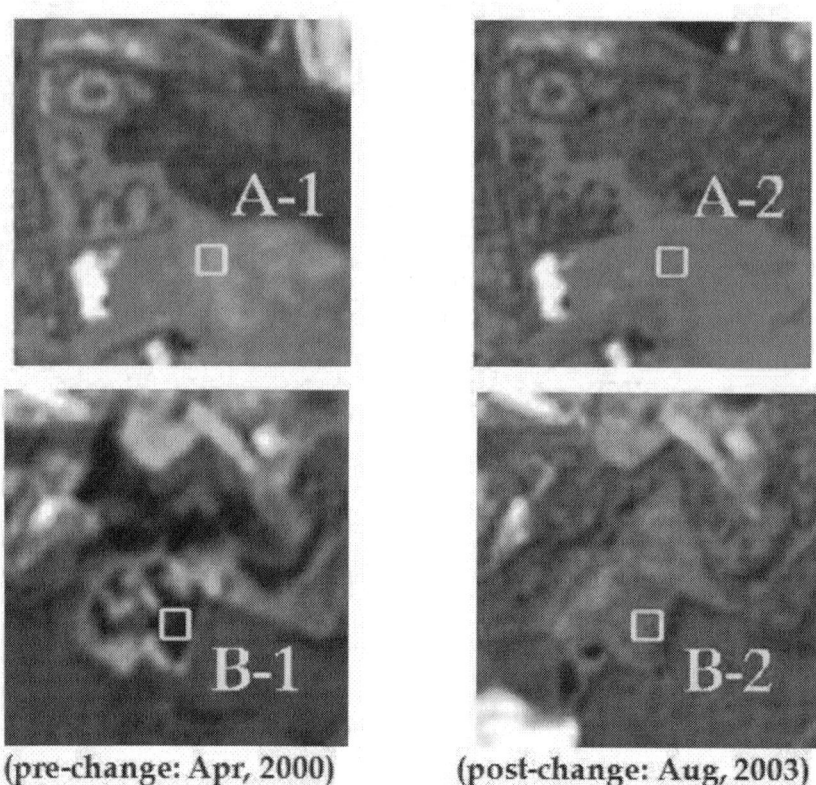

(pre-change: Apr, 2000) (post-change: Aug, 2003)

Figure 2. Particular of two changed areas in an urban environment, located in Maryland, U.S.: CIR visualization (RGB=3N,2,1), ASTER scene of April 9th, 2000 (pre-change: A-1; B-1); CIR visualization (RGB=3N,2,1), ASTER scene of August 24th, 2003 (post-change: A-2; B-2): in the A area (above) a change in land cover from Bare Soil to Vegetation is highlighted in the yellow square, whereas in the B area a change in land cover from dark construction asphalt to paving concrete is highlighted in the yellow square.

Table 1. Normalized Reflectance Difference (NDR) results compared with common Reflectance Difference (RD) results, calculated for particular spots in Figure 2, to show the enhanced discrimination capabilities of NDR.

Area	Land Cover		NIR XE "Spectral range, Near InfraRed (NIR)" Reflectance XE "Reflectance" Response [760-900 nm]		Reflectance XE "Reflectance" Differenc XE "Reflectance Differencing (RD)" e (RD)	Normalized Difference Reflectance (NDR) XE "Normalized Difference Reflectance (NDR)"
Fig. 2	Pre (A-1; B-1)	Post (A-2; B-2)	pre	post		
A	Bare Soil XE "Soil, response"	Grass Vegetati on XE "Vegetat ion"	0.31 1	0.412	0.101	0.14
B	Construction Asphalt	Construc tion Concrete	0.10 3	0.207	0.104	0.335

NORMALIZED DIFFERENCE CHANGE DETYECTION (NDCD)

The further step to exploit the NDR approach defined and described in the previous section is its implementation in the frame of the so called Normalized Difference Change Detection (NDCD) techniques.

The NDCD technique uses the NDR defined in Equation (4.1) as input variables for deriving a land cover/land use change map through the use of one particular change detection method, thus leading to a specific implementation of the NDCD. Out of a range of techniques, such as multi-spectral transforms (e.g. Principal Components Analysis and Minimum Noise Fraction), image classification techniques (both supervised and unsupervised), image segmentation algorithms, Neural Networks or Support Vector Machines, one could be used (Lu et al., 2004).

The possible applications and purposes of this approach are manifold and diverse. In the following we will show the effectiveness of the NDCD for flood mapping. Nevertheless, this approach is a general one and might be applied not only for such mapping purposes, but also for urban growth, burnt areas mapping and other land cover change analyses.

During this work we particularly focused our research on the Minimum Noise Fraction (MNF) (Green et al., 1988, Gianinetto & Villa, 2007) implementation of the normalized difference change detection technique (NDCD-MNF), where the MNF transform is applied to the NDR data to obtain the final change detection map, for the case study analysis of the flood event due to Hurricane Katrina aftermath over the city of New Orleans, in Louisiana, U.S.. The case study and its results will be presented in the next section.

In order to give a demonstration of how the NDR and NDCD approaches work, in the following paragraphs a couple of implemented examples are show, covering a flood hazard mapping case for the monsoonal flood occurred in autumn 2000 and an urban sprawl assessment case for the suburban areas of Washington, U.S..

Flood Hazard, an Example

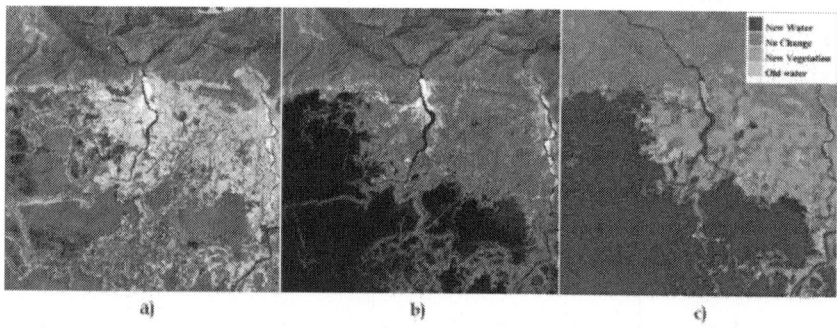

Figure 3. Change Detection for an monsoon flood event, which took place in the Haor region, North-East of Bangladesh: band composition visualization (RGB=7,5,3), ETM+ scene of February 28th, 2000 (a); band composition visualization (RGB=7,5,3), ETM+ scene of October 25th, 2000 (b); Change map derived with Max. Likelihood classification of NDR values.

The first example deal with a change detection application for post-flood analysis, a topic already taken into consideration by previous works of the authors (Gianinetto & Villa, 2006). The inundation event is a monsoon flooding which drawn the North of Bangladesh and North-eastern part of

India in autumn 2000. A pair of Landsat ETM+ scenes covering the Haor region in north-eastern Bangladesh (the pre-event image of February 28th was normalized using post-event image of October 25th as reference) was processed and radiometrically normalized with Pseudo Invariant features (PIFs) selection and linear regression, to produce NDR values as using equation 1.

In order to map surface features changes the couple of images was inspected and regions of change were chosen as ground truth for producing a Maximum Likelihood classification and therefore a map of changed areas, shown in Figure 3. This way, not only the area covered by flooding water could be identified, but also the different vegetation phenological features due to seasonal variations was mapped (Rogan et al., 2002).

Urban Area, an Example

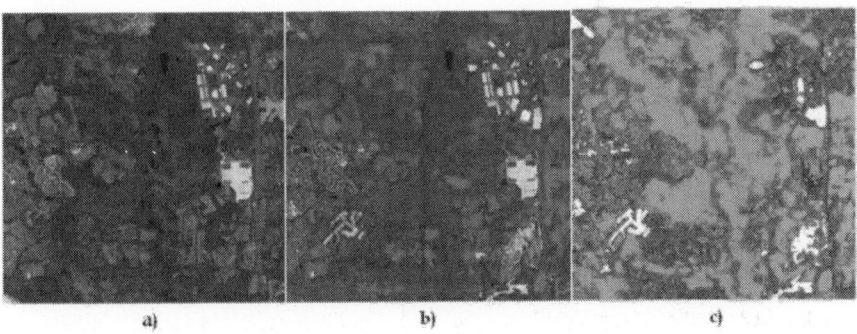

a)   b)   c)

Figure 4. Change Detection for an urban scenario, located in Maryland, U.S., particular of an area of residential and commercial growth in 2000-2003 period: CIR visualization (RGB=3N,2,1), ASTER scene of April 9th, 2000 (a); CIR visualization (RGB=3N,2,1), ASTER scene of August 24th, 2003 (b); Change map derived with ISODATA classification of NDR values. Gray tones represent not changed features, green hues represent increasing vegetation vigour, bright areas represent changed surface cover: mainly newly exposed areas, construction sites and new impervious surfaces.

Another example focuses on a change detection analysis of an urbanized area for urban sprawl and its impact on environment description (Chou et al., 2005). The area covered by Terra ASTER satellite data (VIS, NIR and SWIR subset bands) is located in Maryland, U.S., in the outskirts of

Washington, around 15 kilometres East of the capital's centre: the pre-change image dates back to April 9th, 2000 and was radiometrically normalized using post-change image of August 24th, 2003 as reference, using a linear regression model and PIFs.

After pre-processing and radiometric normalization, the dataset was converted to NDR values, using equation 1, and an unsupervised approach was chosen to classify changes occurred between the two dates (Bruzzone & Prieto, 2000). ISODATA classification was then performed over NDR bands and post classification labelling was utilized to assign to retrieved classes a land cover change significance. The results are displayed in Figure 4 for a small area of detail and class colour code illustrated in caption.

The two examples presented exploit supervised or unsupervised classification of NDR values to produce change maps of a flood event (see Figure 3) or an urban growth situation (see Figure 4); the case studies tests showed a good performance in change areas delineation and identification, as a visual inspection of resulting maps witnesses. It should be pointed out that those example are only representative of a first assessment of NDR approach as an aid to Change Detection; in fact, a thoroughly assessment of the NDCD approach capabilities, together with a comparison with other Change Detection techniques results, will be done in the next section over the complete case study covering Hurricane Katrina struck New Orleans city.

APPLICATION STUDY – FLOOD DAMAGE ASSESSMENT WITH NDCD

Introductive Section

Recent years have seen a tremendous increase in economic and human losses from weather hazards all over the world. Major global climatic alterations are projected to occur during the 21st century and there is great concern about expected negative economic and social consequences resulting from such changes (United Nations, 2007).

Hurricane Katrina was the costliest and one of the deadliest hurricanes in the history of the USA. It was the sixth-strongest Atlantic hurricane ever recorded and the third-strongest land falling U.S. hurricane on record. At its highest intensity, Katrina was a category 5 storm on the Saffir-Simpson scale (Simpson, 1974) with wind speeds of 280 km/h.

The storm made initial landfall at Plaquemines Parish in south-eastern Louisiana on the morning of August 29 2005, and the cities of New Orleans (Louisiana), Mobile (Alabama) and Gulfport (Mississippi) bore the brunt of Katrina's force as it moved inland.

Thanks to the increasing number and observation capabilities of operational remote sensing satellites, remote sensing technology is becoming more and more used for natural hazards monitoring and management, with the great advantage of providing a synoptic vision over a wide area in a short time and in a very cost effective manner (Wang et al., 2002, Brivio et al., 2002, Sanyal & Lu, 2004, Villa & Gianinetto, 2006). In particular, remotely sensed data collected both by radar and optical satellites have been largely used for flood extent evaluation during the last 20 years and now the processing techniques are mature for an operational use (Imhoff et al., 1987, Hess et al., 1995, Frazier et al., 2003, Wang, 2004, Villa & Gianinetto, 2006, Gianinetto & Villa, 2007).

This case study exploits a new method for change detection based on the normalized difference change detection technique (NDCD). The NDCD is a technique which, given an image pair, performs calculations on radiometric normalized reflectance data through the definition of the normalized difference reflectance (NDR) and produces a standardized difference of the reflectance values.

The NDCD was used to detect the damages to the urban area of the city of New Orleans (Louisiana, USA) resulting from the passage of hurricane Katrina. Flood maps were both obtained from the image processing of SPOT-4/HRVIR and Landsat-5/TM imagery, with a suitable spatial resolution for supporting political institutions with a rapid response, effective and prompt decision maker tool.

The maps' accuracy were verified with respect to the inundation maps produced at the Dartmouth Flood Observatory, Dartmouth College (USA). A comparison was also performed between the results of the NDCD technique and that of other standard change detection methods as NIR normalized difference and spectral-temporal minimum noise fraction technique (ST-MNF).

Dataset

Remotely Sensed Dataset

The flooding caused by Hurricane Katrina over the city of New Orleans (29 57' 33" latitude north, 90 03' 36" longitude west) was studied using SPOT-4/HRVIR images supplied by SpotImage and the Centre National d'Etudes Spatiales (CNES) under the Optimising Access to Spot Infrastructure for Science (OASIS) Programme and Landsat-5/TM images made available from the United States Geological Survey's Earth Resources Observation and Science (USGS EROS) through the Hurricane Katrina disaster response project.

The SPOT-4/HRVIR data set was composed of:

- One 20-meters SPOT-4/HRVIR image collected on January 17, 2005 (scene ID 4 601-290 05-01-17 17:03:19 2 I) with orientation angle of 11.5 degree and incidence angle of 19.9 degree left and geocoded in UTM-WGS84 F16N projection. This image was used as pre flood image;
- One 20-meters SPOT-4/HRVIR image collected on September 19, 2005 (Scene ID 4 601-290 05-09-19 16:50:34 1 I) with orientation angle 10.0 degree and incidence angle 0.3 degree right and geocoded in UTM-WGS84 F16N projection. This image was used as post flood image.

The Landsat-5/TM data set was composed of:

- One 30-meters Landsat-5/TM image collected on June 19, 2005 (scene ID 5022039000517010), WRS-2 path 022 row 039, used as pre flood image;
- One 30-meters Landsat-5/TM image collected on September 7, 2005 (scene ID 5022039000525010), WRS-2 path 022 row 039, used as post flood image.

Additional Dataset

For the urban analysis some additional vector maps were used. The 30-meters National Land Cover Database Imperviousness Layer (NLCDIL) raster file representing urbanized and infrastructural features (impervious areas) of the city and surroundings of New Orleans (Yang et al., 2003), made available by USGS through its website (U.S. Geological Survey, 2006) was used for deriving separate mapping for the urban areas only and for the non-urban areas only.

Methodological Approach
Pre-processing
As typical in change detection applications and as envisaged in the earlier part of this work, about pre-processing of data for change analysis, geocoding and atmospheric correction are always needed. For this purposes the satellite data were first georeferenced in the UTM-WGS84 projection, using reference data. Original at-sensor radiance data were atmospherically corrected using a low resolution Radiative Transfer Code, combined with aerosol retrieval based on band reflectance ratios and with adjacency correction of path radiance (Berk et al., 1999, Vermote et al., 1997).

A further step is the radiometric normalization of multispectral data, carried out using a parabolic parametric model:

$$Rjnorm = aj(Rjraw)bj$$

(2)

where:

R^{norm} = normalized reflectance
R^{raw} = input reflectance of the slave image
a = multiplicative coefficient of the parametric model
b = exponential coefficient of the parametric model
j = spectral band number

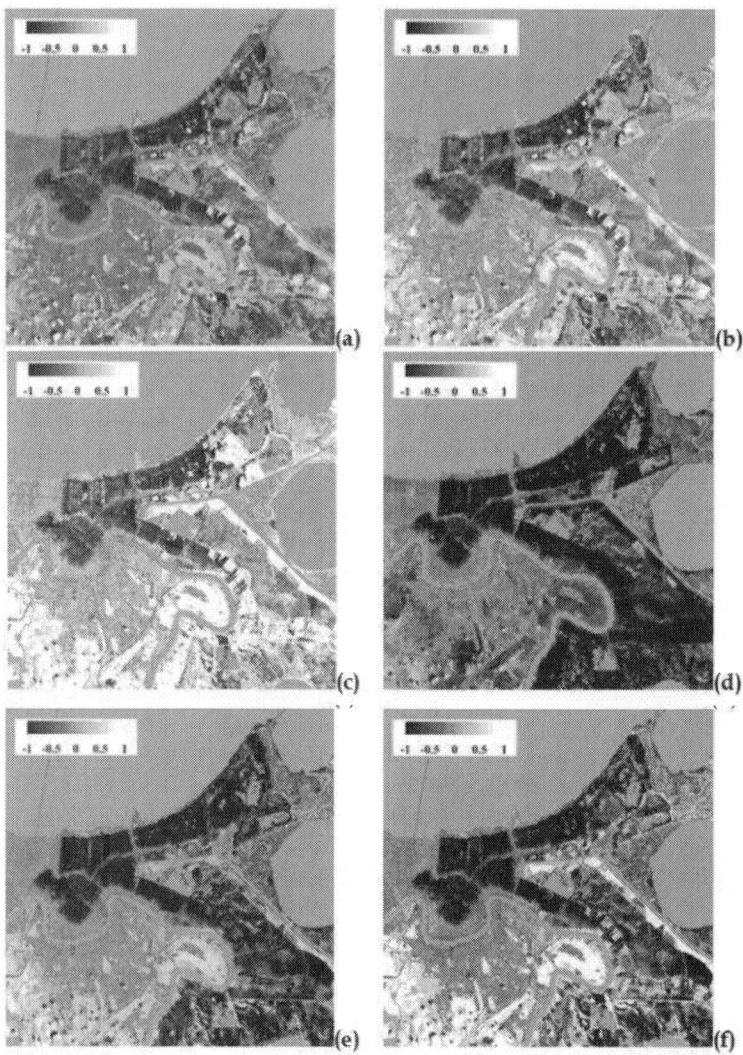

Figure 5. Example of normalized difference reflectance (NDR) calculated for the Landsat-5/TM dataset. (a) Spectral band nr.1; (b) Spectral band nr.2; (c) Spectral band nr.3; (d) Spectral band nr.4; (e) Spectral band nr.5; (f) Spectral band nr.7.

The radiometric normalization of reflectance data was performed using a parametric parabolic model based on equation 2 through standard linearized least square matching based on a parametric model and an iteration approach to solution of the linearized basic observation equation. The transformation coefficients were computed using standard linearized least squares matching, through an iteration approach to solution of the linearized basic observation equation.

Radiometrically normalized data were used for calculating NDR values for both Landsat-5/TM and SPOT-4/HRVIR data, using the approach described in the previous sections and calculated with equation (4.1). The NDR values were finally used as inputs for Minimum Noise Fraction (MNF) transform, thus structuring the implementation of the NDCD-MNF technique for change mapping and flooded area delineation.

Mapping Hurricane Katrina's aftermaths in New Orleans
The widespread destruction in New Orleans was mapped using the NDCD-MNF technique. The SPOT-4/HRVIR and Landsat-5/TM images were first radiometrically normalized using the parametric model of equation (2.2) and the NDR were computed using equation (4.1). Following, to the multi-spectral NDR values it was applied the MNF transform, generating the normalized difference reflectance-Minimum Noise Fraction (NDR-MNF) components.

From all the NDR-MNF components generated, only the first and the second were retained, being the most representative of a good identification of water related land cover. By visual interpretation of the post flood images, the final selection of the best representative NDR-MNF component (component nr.1 or component nr.2) was carried out and the final mapping was realized by using an adaptive threshold. Figure 7 shows the NDR-MNF component nr.2 for SPOT-4/HRVIR (Figure 7a) and Landsat-5/TM (Figure 7b), subsequently used for the mapping.

a)   b)

Figure 6. Normalized Difference Reflectance-Minimum Noise Fraction (NDR-MNF) component nr.2 used for the mapping. (left) SPOT-4/HRVIR; (right) Landsat-5/TM.

The criteria used for the threshold selection was based on the detection of the maximum separability interval between flooded and non-flooded areas. For the SPOT-4/HRVIR and Landsat-5/TM image data some samples of the selected NDCD-MNF component were independently extracted, belonging to the urban and non-urban land cover classes, both for the flooded and non-flooded areas.

For the Landsat-5/TM data set, 900 pixels were selected for the urbanized areas (400 pixels in flooded area and 500 in non-flooded area), covering nearly 0.2% of the total urbanized areas, while 1,200 pixels were selected for the non-urbanized areas (700 pixels in flooded area and 500 in non-flooded area), covering nearly 0.2% of the total non-urbanized areas.

For the SPOT-4/HRVIR data set, 1,500 pixels were selected for the urbanized areas (600 pixels in flooded area and 900 in non-flooded area), covering nearly 0.15% of the total urbanized areas, while 2,000 pixels were selected for the non-urbanized areas (1,200 pixels in flooded area and 800 in non-flooded area), covering nearly 0.15% of the total non-urbanized areas.

For all these samples, the first and second-order statistics were computed and the maximum separability interval between the flooded and non-flooded areas was identified by testing different threshold values belonging to the interval; finally the global flood maps were produced.

Next, using the USGS's NLCDIL as supplementary input data, three other products were generated from the SPOT-4/HRVIR and the Landsat-5/TM data sets: i) a flood map for the 'urban areas only'; ii) a flood map for the 'non-urban areas only'; and iii) a 'fused' flood map:

1. The flood map for the urban areas only was built using the non-impervious surface layer of the NLCDIL as mask for excluding from the processing all the image pixels collected on non-urban areas;

2. The flood map for the non-urban areas only was built using the impervious surface layer of the NLCDIL as mask for excluding from the processing all the image pixels collected on urban areas;

3. The fused flood map was built fusing together the results previously obtained for the urban areas only and the non-urban areas only. This processing returned a product comparable to the global flood map above described, but it has proven more accurate.

To boost the spatial coherency and homogeneity of the final mapping, all the flood maps were refined with classical segmentation and clumping techniques.

Performance Evaluation and Comparison to Other Techniques

The accuracies of all the maps produced with the NDCD-MNF technique were verified using as ground truth the flood extension map of the city of New Orleans (Figure 8) produced at the Dartmouth Flood Observatory (Dartmouth College, USA) and provided by courtesy of Prof. G.R. Brakenridge and Dr. E. Anderson (Dartmouth College, USA).

The potentialities and performances of the NDCD technique for flood mapping were also compared to following standard change detection methods characterized by different complexity:

- Change detection based on the near-infrared normalized difference (Hayes and Sader, 2001);
- Spectral-Temporal Minimum Noise Fraction (ST-MNF) technique previously developed by authors for flood mapping (Gianinetto & Villa, 2007).

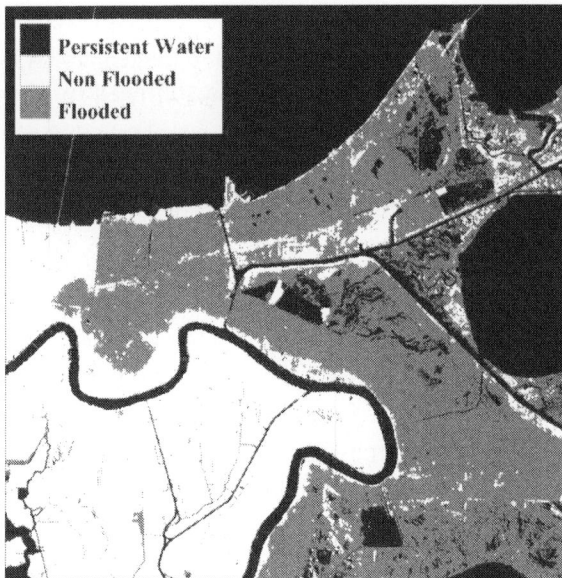

Figure 7. Meters resolution raster image derived from the vector flood extension map produced at the Dartmouth Flood Observatory (Dartmouth College, USA).

NIR Normalized Difference Change Detection

The simplest change detection technique used to compare the results obtained using the NDCD-MNF was based on the NIR normalized difference (Hayes and Sader, 2001). Using only the infrared band (TM4 for Landsat-5/TM and XS3 for SPOT-4/HRVIR) it was produced a flood map by thresholding the normalized difference between the post-flood and pre-flood images.

Similarly to the processing carried on with the NDCD-MNF, also using the NIR normalized difference it were separately calculated: i) a flood map for the urban areas only, ii) a the flood map for the non-urban areas only, and iii) a global flood map, both for the SPOT-4/HRVIR and the Landsat-5/TM data set.

Spectral-Temporal Minimum Noise Fraction technique

Another term of comparison for the NDCD-MNF method was the ST-MNF technique previously developed by the authors (Gianinetto & Villa, 2006, Gianinetto & Villa, 2007).

In this case only the global flood maps were generated for both the Landsat-5/TM and SPOT-4/HRVIR data set by processing together both the impervious and non-impervious land cover features. Starting from the pre-processed normalized images, a synthetic n-band file (with n=8 for SPOT-4/HRVIR and n=12 for Landsat-5/TM) was created including first the reflective bands of the pre flood scene followed by the homologous bands of the post flood scene, stacked together. To this Spectral-Temporal merging it was applied the MNF transform and a thresholding to derive the flood extension map, whereas a complete description of the ST-MNF technique can be found in (Gianinetto & Villa, 2007).

Results and Discussion

Sampling for Accuracy Assessment

The testing samples used for the accuracy assessment of the flood maps were selected following a stratified random sampling approach over the datasets. In detail, accuracy test samples were collected as:

a.   For the SPOT-4/HRVIR data set:

b.   For the Landsat-5/TM data set:

Mapping Accuracy Using the NDCD-MNF Technique

Flood maps for the 'urban areas only' and for the 'non-urban areas only', along with a 'fused' flood map were obtained by thresholding the NDR-MNF component nr.1.

Regarding the flood mapping in the urban areas only, the data processing performed on the Landsat-5/TM imagery led to higher accuracy than those performed on the SPOT-4/HRVIR imagery (Table 2,Figure 9). For the former it was obtained a best Overall Accuracy (OA) of 92.05% and a kappa coefficient (K) of 0.83, while for the latter the results gave an OA of 86.37% and a K of 0.72.

The threshold selection was not a critical issue for the Landsat-5/TM data, while for the SPOT-4/HRVIR data, approaching to the upper (positive) limit of the separability interval the accuracy became worse (OA=72.30, K=0.48). In any case, regardless the threshold value selection, the mapping based on the Landsat-5/TM images was always superior to those based on the SPOT-4/HRVIR images.

On the contrary, with respect to the flood mapping in the 'non-urban areas only', the data processing performed on the Landsat-5/TM imagery led to lower accuracy than those performed on the SPOT-4/HRVIR imagery (Table 3, Figure 10). For the former it was obtained a best OA of 75.70% and a K of 0.49, while for the latter the results gave an OA of 86.31% and a K of 0.71.

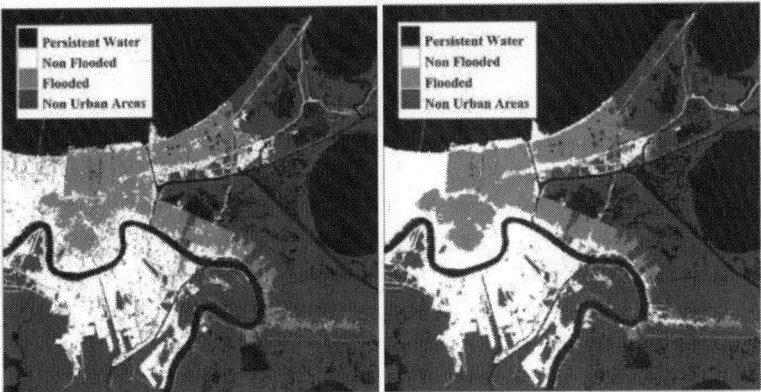

Figure 8. New Orleans (Louisiana). Flood mapping in the urban areas only using the NDCD-MNF technique. (left) Derived from SPOT-4/HRVIR data; (right) Derived from Landsat-5/TM data.

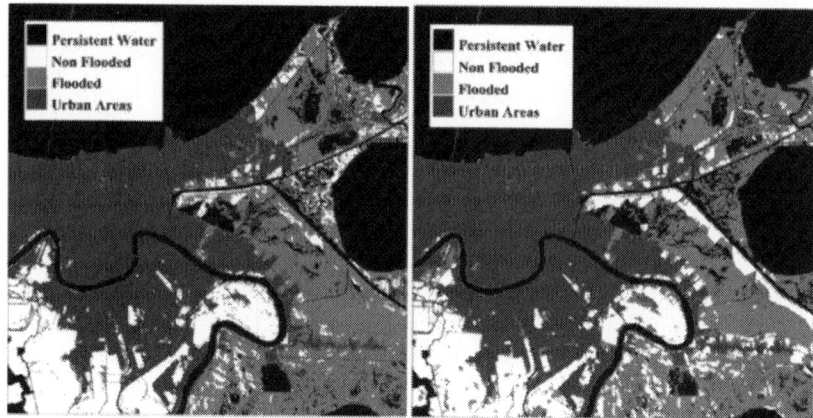

Figure 9. New Orleans (Louisiana). Flood mapping in the non-urban areas only using the NDCD -MNF technique. (left) Derived from SPOT-4/HRVIR data; (right) Derived from Landsat-5/TM data.

The threshold selection was not a critical issue both for the Landsat-5/TM and for the SPOT-4/HRVIR data. In any case, regardless of the threshold value selection, this time the mapping based on the SPOT-4/HRVIR images was always superior to those based on the Landsat-5/TM images.

Table 2. Values in bold indicate the best accuracy.

	Threshold XE "Threshold selection" value	Overall Accuracy XE "Accuracy, Overall Accuracy (OA)" * (%)	K coeff icient
SPOT- XE "Satellite, SPOT" 4 /HRVIR XE "Sensor, HRVIR" data set	-2	85.72	0.7
	-1.5	86.37	0.72
	-1	84.96	0.69
	0	72.3	0.48
Landsat- XE "Satellite, Landsat" 5/TM XE "Sensor, TM" data set	-1	89.69	0.79
	0	90.52	0.8
	1	92.05	0.83
	2	91.16	0.81

Flood mapping in the 'urban areas only' using the NDCD-MNF technique. Threshold selection and mapping accuracy

Table 3. Values in bold indicate the best accuracy.

	Threshold XE "Threshold selection" value	Overall Accuracy XE "Accuracy, Overall Accuracy (OA)" * (%)	K coefficient
SPOT- XE "Satellite, SPOT" 4 /HRVIR XE "Sensor, HRVIR" data set	0	85.56	0.69
	0.5	86.31	0.71
	1	86.17	0.7
	2	84	0.65
Landsat- XE "Satellite, Landsat" 5/TM XE "Sensor, TM" data set	-1.5	75.59	0.49
	-1	75.7	0.49
	-0.5	75.11	0.48
	0	74.34	0.47

Flood mapping in the 'non-urban areas only' using the NDCD-MNF technique. Threshold selection and mapping accuracy.

The reason of the poor mapping in non-urban areas using the Landsat-5/TM dataset may be found in the closeness of the post-flood image (September 7, 2005) to Katrina landfall (August 29, 2005). In the Landsat-5/TM post-flood image many wet areas (rain-washed), mainly located in non urbanized areas (impervious surfaces), were incorrectly detected as flooded. This phenomena was not observed in the SPOT-4/HRVIR post-flood image because of the longer time elapsed from the passage of Katrina (September 19, 2005).

A global 'fused' flood map was obtained by fusing together of the urbanized and non-urbanized flood maps separately computed with the NDCD-MNF technique (Figure 11). In this case, the comparison of global results from the Landsat-5/TM (OA=84.03% and K=0.68) and SPOT-4/HRVIR (OA=86.36% and K=0.73) data processing are similar, with a little advantage for the SPOT-based mapping. This result is justified by the non homogeneous accuracy of the Landsat-based mapping in urban and non urban areas, as previous described (Table 2 and Table 3).

A simpler and less computational expensive global flood map was derived by processing together both the urban and non-urban areas in a single step (Figure 12). Differently to the previous cases, the NDR-MNF component nr.2 was used this time for the thresholding of both the dataset.

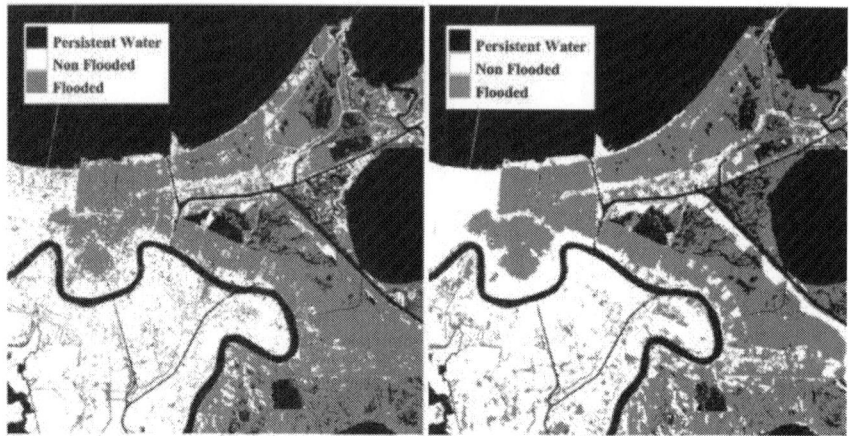

Figure 10. New Orleans (Louisiana). Global 'fused' mapping using the NDCD-MNF technique. (left) Derived from SPOT-4/HRVIR data; (right) Derived from Landsat-5/TM data.

Figure 11. New Orleans (Louisiana). Global flood map using the NDCD-MNF technique. (left) Derived from SPOT-4/HRVIR data; (right) Derived from Landsat-5/TM data.

Again the Landsat-5/TM data processing led to lower accuracy (OA=77.32% and K=0.54) than the SPOT-4/HRVIR data processing (OA=83.20% and K=0.67), when compared to ground truth data, mainly due to its lower accuracy in the non-urban areas (Table 4).

Table 4. Values in bold indicate the best accuracy.

	Threshold XE "Threshold selection" value	Overall Accuracy XE "Accuracy, Overall Accuracy (OA)" * (%)	K coefficient
SPOT- XE "Satellite, SPOT" 4 /HRVIR XE "Sensor, HRVIR" data set	-2	66.11	0.28
	-1	82.02	0.63
	-0.5	83.2	0.67
	0	82.24	0.65
	1	75.92	0.53
Landsat- XE "Satellite, Landsat" 5/TM XE "Sensor, TM" data set	0	76.6	0.53
	1	77.32	0.54
	2	75.81	0.51
	3	72.5	0.45

Global flood mapping using the NDCD-MNF technique when processing together both the impervious and non-impervious surfaces. Threshold selection and mapping accuracy

This time, for the SPOT-4/HRVIR dataset both the NDR-MNF component selection and the threshold selection are critical and the accuracy largely depends upon their correct identification. Regarding the NDR-MNF component selection, when using component 1 instead of component 2, as in previous cases, for the Landsat-5/TM data processing we had a little decrease in the mapping accuracy (from 77.32% to 75.50% for the OA), while for the SPOT-4/HRVIR data processing a greater decrease in the mapping accuracy was observed (from 83.20% to 67.74% for the OA).

The foremost advantage of this single step global mapping is that no urban mask is required, so no a priori information is needed to separate impervious from non impervious surfaces. On the other hand, the mapping accuracy is always worse when compared to the global 'fused' map obtained by fusing together of the urbanized and non-urbanized flood maps separately computed (Figure 11). For the SPOT-4/HRVIR data it was observed a decrease in the OA from 86.36% to 83.20% and a decrease in the K from 0.73 to 0.67, while for the Landsat-5/TM data it was observed a larger decrease both in the OA from 84.36% to 77.32% and in the K from 0.68 to 0.54 (Table 4).

Comparing the Mapping Accuracy of the NDCD-MNF to the NIR-normalized Difference Change Detection
Using the NIR-normalized difference change detection it was generated a global flood map by processing together both the urban and the non-urban areas (Figure 13) and a global 'fused' flood map by processing separately the urban and the non-urban areas (Figure 14).

This technique seems to be insensitive to both the data set used (SPOT-4/HRVIR or Landsat-5/TM) and to the data processing adopted (global or 'fused' map), leading to an OA between 81.66% and 82.75% and a K between 0.63 and 0.66. Table 5 shows a summary of results.

When comparing the NDCD-MNF to the NIR-normalized difference change detection it emerged the superiority of the former in all the 'fused' products (OA=86.36 and K=0.73 for the SPOT-4/HRVIR data set and OA=84.36 and K=0.68 for the Landsat-5/TM data set) and a better accuracy for the latter with respect to only the Landsat-5/TM global flood map (OA=82.03% and K=0.64).

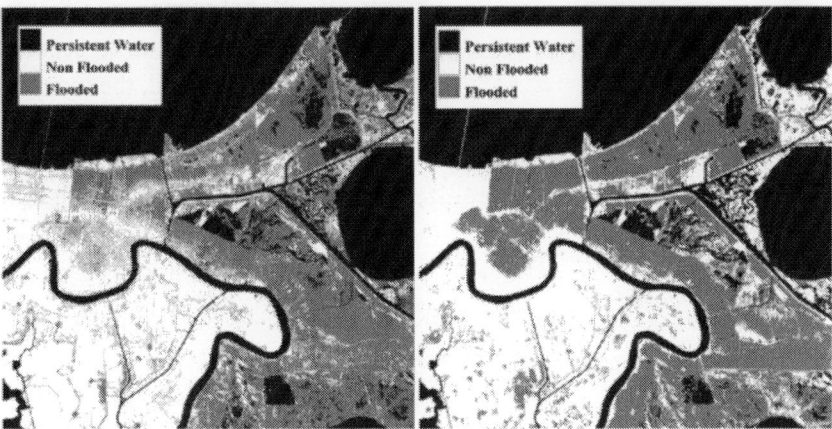

Figure 12. New Orleans (Louisiana). Global flood mapping using the NIR-normalized difference technique. (left) Derived from SPOT-4/HRVIR; (right) Derived from Landsat-5/TM.

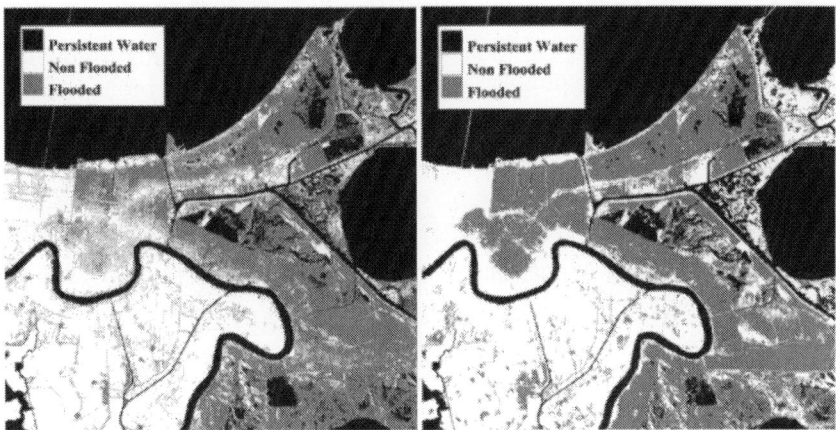

Figure 13. New Orleans (Louisiana). Global 'fused' map using the NIR-normalized difference technique. (left) Derived from SPOT-4/HRVIR; (right) Derived from Landsat-5/TM..

Table 5. Values in bold indicate the best accuracy. Global flood mapping using the NIR-normalized difference change detection technique. Threshold selection and mapping accuracy

	Threshold XE "Threshold selection" Value	Overall Accuracy XE "Accuracy, Overall Accuracy (OA)" * (%)	K coefficient
SPOT- XE "Satellite, SPOT" 4 /HRVIR XE "Sensor, HRVIR" data set	-0.1	79.13	0.59
	-0.075	81.07	0.62
	-0.05	81.73	0.63
	-0.025	80.83	0.61
Landsat- XE "Satellite, Landsat" 5/TM XE "Sensor, TM" data set	-0.35	79.92	0.6
	-0.3	82.03	0.64
	-0.25	81.95	0.64
	-0.2	80.16	0.6

Comparing the Mapping Accuracy of the NDCD-MNF to the Spectral-temporal Minimum Noise Fraction Technique

Using the ST-MNF technique it was generated a global flood map by processing together both the urban and the non-urban areas on the basis of the MNF component nr. 1 for both the data set (Figure 15).

The threshold selection here revealed to be not a critical issue for the mapping accuracy. By using the ST-MNF technique, the Landsat-5/TM data processing led to higher accuracy (OA=90.17% and K=0.80) than the SPOT-4/HRVIR data processing (OA=81.87% and K=0.63) when compared to ground truth data. Table 6 shows a summary of results.

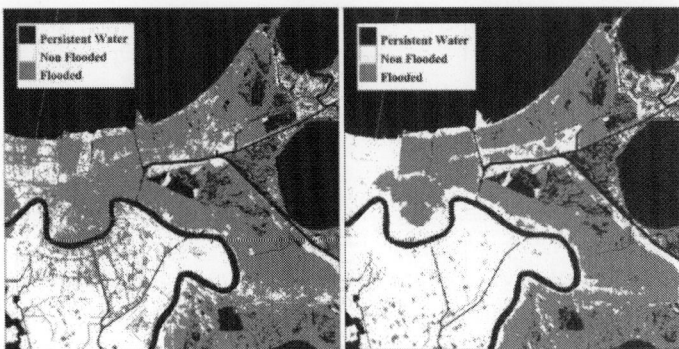

Figure 14. New Orleans (Louisiana). Global damages mapping using the ST-MNF technique. (left) Derived from SPOT-4/HRVIR data; (right) Derived from Landsat-5/TM data.

A comparison between the NDCD-MNF and the ST-MNF shows that the former always performed better on the SPOT-4/HRVIR data set, regardless the data processing used for the global mapping used (with or without urban areas masking). While with respect to the Landsat-5/TM data set, results are more difficult to analyse. Looking at the OA only it seems that the ST-MNF led to higher accuracy (OA=90.17 for ST-MNF and OA=84.36 for NDCD-MNF), but the K score of the NDCD-MNF is higher for the 'fused' map (K=0.68 for NDCD-MNF and K=0.63 for ST-MNF). So it is difficult to say which performed better with respect to the overall situation.

Table 6. Values in bold indicate the best accuracy.

	Threshold XE "Threshold selection" value	Overall Accuracy XE "Accuracy, Overall Accuracy (OA)" * (%)	K coefficient
SPOT- XE "Satellite, SPOT" 4 /HRVIR XE "Sensor, HRVIR" data set	-1	81.76	0.63
	-0.5	81.87	0.63
	0	81.43	0.62
	1	79.33	0.58
Landsat- XE "Satellite, Landsat" 5/TM XE "Sensor, TM" data set	-1	89.37	0.78
	0	89.79	0.79
	0.5	90.12	0.8
	1	90.17	0.8
	2	89.71	0.79

Global flood mapping using the ST-MNF technique. Threshold selection and mapping accuracy.

Summary and Conclusions

This case study tested the normalized difference change detection technique effectiveness for change mapping, also in comparison with other literature methodologies, starting from the processing of the normalized difference reflectance data.

The radiometric normalization of data influenced the accuracy of the mapping. A parametric normalization with coefficients calculated with standard linearized least squares adjustment and iterative solution was found a better solution than a standard linear normalization, and thus adopted. However, the general definition of the NDCD leaves the possibility to develop processing techniques based on different radiometric normalization schemes.

Using its MNF implementation, the NDCD technique was used for mapping and evaluating the havoc on the city of New Orleans (Louisiana, USA) wreaked by Hurricane Katrina landfall in August 2005, using both a SPOT-4/HRVIR and a Landsat-5/TM data set.

As a term of comparison for evaluating the potentialities and performances of the NDCD-MNF technique, several other standard change detection methods have been tested: from the simple NIR normalized difference to the more complex Spectral-Temporal Minimum Noise Fraction technique.

Comparing the global mapping accuracy when using the SPOT-4/HRVIR data, the NDCD-MNF technique always leaded to better results than all the others methods here taken into consideration. Moreover, results were better when processing separately the urban and the non-urban areas in the so called 'fused' product.

With regards to the Landsat-5/TM data, the NDCD-MNF technique poorly performed in the non-urban areas (probably due to the closeness of the post-flood image to Katrina landfall), thus affecting the final global mapping. However, with respect to the only urban areas, which may be of major interest in most cases, the NDCD-MNF always performed better.

Finally, regarding the threshold selection, a number of studies [Yuan et al., 1998; Chen et al., 2003] have pointed out that a major weakness of all spectral change detection approaches is that the selection of a minimum threshold to signify change is often arbitrary [Warner, 2005]. For example, a threshold value of two standard deviations above the mean is sometimes selected (Sohl, 1999). To address this problem, some studies used a noise model to select the threshold (Dwyer et al., 1996). As an alternative, other studies have developed a systematic method using training data (Chen et al., 2003). In their approach, areas of change are digitized, as well as a surrounding window of no change. These training areas are then classified into 'change' and 'no change' classes, using a small number of arbitrarily chosen thresholds spread over a wide range of possible values. Based on the accuracies of these classifications, the range of thresholds is narrowed successively to focus on the region where the accuracy is highest. In this iterative fashion, an optimal threshold is selected.

The accuracy gained through NDR approach derived changes mapping have been proven very satisfying most of the times, with Overall Accuracies percentage figures ranging from 80% to over 90% of correctness, that is to say error percentage in change mapping around 10% which is to be considered as a really good result for analysis performed and based only on remote sensing satellite data.

Nevertheless, more testing and a more fine tuning of the processing chain can be implemented and done in the future, including not yet explored land cover change application fields, and the good results achieved until now are a great encouragement to continue on the path already traced with the works described in this chapter.

REFERENCES

1. A. Berk, G. P. Anderson, L. S. Bernstein, P. K. Acharya, H. Dothe, M. W. Matthew, S. M. Adler-Golden, J. H. Chetwynd, S. C. Richtsmeier, B. Pukall, C. L. Allred, L. S. Jeong, M. L. Hoke, 1999 MODTRAN4 Radiative Transfer Modeling for Atmospheric Correction, SPIE Proceedings, Optical Spectroscopic Techniques and Instrumentation for Atmospheric and Space Research III, 3756 348 353

2. P. A. Brivio, R. Colombo, M. Maggi, R. Tomasoni, 2002 Integration of remote sensing data and GIS for accurate mapping of flooded areas, International Journal of Remote Sensing, 23(3), pp. 429-441.

3. L. Bruzzone, Prieto. D. Fernàndez, 2000 Automatic Analysis of the Difference Image for Unsupervised Change Detection, IEEE

Transactions on Geoscience and Remote Sensing, vol. 38, no. 3, pp.1171-1182.

4. J. Chen, P. Gong, P. Shi, 2003 Land-use/land-cover change detection using improved change-vector analysis , Photogrammetric Engineering and Remote Sensing, 69, 369 379 .

5. T. Y. Chou, T. C. Lei, S. Wan, L. S. Yang, 2005 Spatial knowledge databases as applied to the detection of changes in urban land use, International Journal of Remote Sensing, 26 14 3047 3068 .

6. P. Coppin, M. E. Bauer, 1994 Digital Processing of Multitemporal Landsat TM Imagery to Optimize Extraction of Forest Cover Change Features, IEEE Transactions on Geoscience and Remote Sensing, 32(4), pp. 918-927.

7. P. Coppin, K. Nackaerts, L. Queen, K. Brewer, 2001 Operational Monitoring of Green Biomass Change for Forest Management, Photogrammetric Engineering & Remote Sensing, 67(5), 603 611 .

8. P. Coppin, I. Jonckheere, K. Nackaerts, B. Muys, E. Lambin, 2004 Digital change detection methods in ecosystem monitoring: a review, International Journal of Remote Sensing, 25(9), 1565 1596 .

9. J. L. Dwyer, K. L. Sayler, G. J. Zylstra, 1996 Landsat pathfinder data sets for landscape change analysis. Proceedings of the 1996 IEEE International Geoscience and Remote Sensing Symposium, IGARSS 1996, Lincoln, Nebraska, pp. 547-550.

10. P. Frazier, K. Page, J. Louis, S. Briggs, A. I. Robertson, 2003 Relating wetland inundation to river flow using Landsat TM data, International Journal of Remote Sensing, 24(19), pp. 3755-3770.

11. M. Gianinetto, P. Villa, 2006 Monsoon Flooding Response: a Multi-scale Approach to Water-extent Change Detection, The International Archive of the Photogrammetry, Remote Sensing and Spatial Information Sciences, XXXVI(7), Enschede, the Netherlands, pp. 128-133.

12. M. Gianinetto, P. Villa, 2007 Rapid Response Flood Assessment Using Minimun Noise Fraction And Composed Spline Interpolation, IEEE Transactions on Geoscience and Remote, 45(10), pp. 3204-3211.

13. A. A. Green, M. Berman, P. Switzer, M. D. Craig, 1988 A transformation for ordering multispectral data in terms of image quality with implications for noise removal, IEEE Transactions on Geoscience and Remote Sensing, 26(1), 65 74 .

14. D. J. Hayes, S. A. Sader, 2001 Comparison of Change-Detection Techniques for Monitoring Tropical Forest Clearing and Vegetation Regrowth in a Time Series, Photogrammetric Engineering and Remote Sensing, 67(9), pp. 1067-1075.

15. L. L. Hess, J. M. Melack, S. Filoso, Y. Wang, 1995 Realtime mapping of inundation on the Amazon floodplain with the SIR-C/X-SAR synthetic aperture radar, IEEE Transactions on Geoscience and Remote Sensing, 33, pp. 896-904.

16. M. L. Imhoff, C. Vermillon, M. H. Story, A. M. Choudhury, A. Gafoor, 1987 Monsoon flood boundary delineation and damage assessment using spaceborne imaging radar and Landsat data, Photogrammetric Engineering and Remote Sensing, 4, pp. 405-413.

17. D. Lu, P. Mausel, E. Brondizio, E. Moran, 2004 Change detection techniques, International Journal of Remote Sensing, 25, 2365 2407 .

18. J. Rogan, J. Franklin, D. A. Roberts, 2002 A comparison of methods for monitoring multitemporal vegetation change using Thematic Mapper imagery, Remote Sensing of Environment, 80 143 156 .

19. J. Sanyal, X. Lu, X. , 2004 Application of the Remote Sensing in Flood Management with Special Reference to Monsoon Asia: a Review, Natural Hazards, 33, pp. 283-301.

20. R. H. Simpson, 1974 The hurricane disaster-potential scale, Weatherwise, 27, 169

21. A. Singh, 1989 Digital change detection techniques using remotely-sensed data, International Journal of Remote Sensing, 10(6), 989 1003.

22. T. L. Sohl, 1999 Change analysis in the United Arab Emirates: an investigation of techniques, Photogrammetric Engineering and Remote Sensing, 65, 475 484 .

23. GeologicalU. S.Survey 2001 National Land Cover Database 2001 (NLCD 2001), digital resource available online at:www.mrlc.gov/mrlc2k_nlcd.asp.

24. United Nations, 2007 Evidence is now'unequivocal' that humans are causing global warming, UN report.

25. E. Vermote, D. Tanré, J. L. Deuzé, M. Herman, J. J. Morcrette, 1997 Second Simulation of the Satellite Signal in the Solar Spectrum, 6S: An Overview, IEEE Transactions on Geoscience and Remote Sensing, 35(3), 675 686 .

26. P. Villa, M. Gianinetto, 2006 Multispectral transform and Spline Interpolation for Mapping Flood Damages, Proceedings of the 2006 IEEE International Geoscience and Remote Sensing Symposium, IGARSS 2006, Denver, U.S., pp. 275-279.

27. P. Villa, G. Lechi, 2007 Normalized Difference Reflectance : An Approach to Quantitative Change Detection, Proceedings of the 2007 IEEE International Geoscience and Remote Sensing Symposium, IGARSS 2007, Barcelona, Spain, pp. 2366-2369.

28. Y. Wang, J. D. Colby, K. A. Mulcahy, 2002 An efficient method for mapping flood extent in a coastal flood plain using Landsat TM and DEM data, International Journal of Remote Sensing, 23(18), pp. 3681-3696.

29. Y. Wang, 2004 Using Landsat 7 TM data acquired days after a flood event to delineate the maximum flood extent on a coastal floodplain, International Journal of Remote Sensing, 25(5), pp. 959-974.

30. L. Yang, C. Huang, C. G. Homer, B. K. Wylie, M. J. Coan, 2003 An approach for mapping large-area impervious surfaces: synergistic use of Landsat -7 ETM+ and high spatial resolution imagery, Canadian Journal of Remote Sensing, 29(2), pp. 230-240.

CITATION

Paolo Villa, Giovanmaria Lechi and Mario A. Gomarasca (2009). Multivariate Differencing Techniques for Land Cover Change Detection: the Normalized Difference Reflectance Approach, Geoscience and Remote Sensing, Pei-Gee Peter Ho (Ed.), ISBN: 978-953-307-003-2, InTech, DOI: 10.5772/8312.

Index